Revisiting the *Origin of Species*

Contemporary interest in Darwin rises from a general ideal of what Darwin's books ought to contain: a theory of transformation of species by natural selection. However, a reader opening Darwin's masterpiece, *On the Origin of Species*, today may be struck by the fact that this "selectionist" view does not deliver the key to many aspects of the book. Without contesting the importance of natural selection to Darwinism, much less supposing that a fully-formed "Darwinism" stepped out of Darwin's head in 1859, this innovative volume aims to return to the text of the *Origin* itself.

Revisiting the Origin of Species focuses on Darwin as theorising on the origin of variations; showing that Darwin himself was never a pan-selectionist (in contrast to some of his followers) but was concerned with "other means of modification" (which makes him an evolutionary pluralist). Furthermore, in contrast to common textbook presentations of "Darwinism", Hoquet stresses the fact that *On the Origin of Species* can lend itself to several contradictory interpretations. Thus, this volume identifies where rival interpretations have taken root; to unearth the ambiguities readers of Darwin have latched onto as they have produced a myriad of Darwinian legacies, each more or less faithful enough to the originator's thought.

Emphasising the historical features, complexities and intricacies of Darwin's argument, *Revisiting the* Origin of Species can be used by any lay readers opening Darwin's *On the Origin of Species*. This volume will also appeal to students and researchers interested in areas such as Evolution, Natural Selection, Scientific Translations and Origins of Life.

Thierry Hoquet is Professor of Philosophy of Science in the Department of Philosophy at Paris Nanterre University, France.

History and Philosophy of Biology
Series Editor: Rasmus Grønfeldt Winther
rgw@ucsc.edu | www.rgwinther.com

This series explores significant developments in the life sciences from historical and philosophical perspectives. Historical episodes include Aristotelian biology, Greek and Islamic biology and medicine, Renaissance biology, natural history, Darwinian evolution, Nineteenth-century physiology and cell theory, Twentieth-century genetics, ecology, and systematics, and the biological theories and practices of non-Western perspectives. Philosophical topics include individuality, reductionism and holism, fitness, levels of selection, mechanism and teleology, and the nature-nurture debates, as well as explanation, confirmation, inference, experiment, scientific practice, and models and theories vis-à-vis the biological sciences.

Authors are also invited to inquire into the "and" of this series. How has, does, and will the history of biology impact philosophical understandings of life? How can philosophy help us analyze the historical contingency of, and structural constraints on, scientific knowledge about biological processes and systems? In probing the interweaving of history and philosophy of biology, scholarly investigation could usefully turn to values, power, and potential future uses and abuses of biological knowledge.

The scientific scope of the series includes evolutionary theory, environmental sciences, genomics, molecular biology, systems biology, biotechnology, biomedicine, race and ethnicity, and sex and gender. These areas of the biological sciences are not silos, and tracking their impact on other sciences such as psychology, economics, and sociology, and the behavioral and human sciences more generally, is also within the purview of this series.

Rasmus Grønfeldt Winther is Associate Professor of Philosophy at the University of California, Santa Cruz (UCSC), and Visiting Scholar of Philosophy at Stanford University (2015–2016). He works in the philosophy of science and philosophy of biology and has strong interests in metaphysics, epistemology, and political philosophy, in addition to cartography and GIS, cosmology and particle physics, psychological and cognitive science, and science in general. Recent publications include "The Structure of Scientific Theories", *The Stanford Encyclopaedia of Philosophy* and "Race and Biology", *The Routledge Companion to the Philosophy of Race*. His book with University of Chicago Press, *When Maps Become the World*, is forthcoming.

Modelling Evolution
A New Dynamic Account
Eugene Earnshaw

Revisiting the *Origin of Species*
The Other Darwins
Thierry Hoquet

Revisiting the *Origin of Species*

The Other Darwins

Thierry Hoquet

Routledge
Taylor & Francis Group

LONDON AND NEW YORK

First published 2018
by Routledge
2 Park Square, Milton Park, Abingdon, Oxon OX14 4RN

and by Routledge
711 Third Avenue, New York, NY 10017

Routledge is an imprint of the Taylor and Francis Group, an informa business

British Library Cataloguing-in-Publication Data
A catalogue record for this book is available from the British Library

Library of Congress Cataloging-in-Publication Data
A catalog record has been requested for this book

ISBN: 978-1-138-60715-6 (hbk)
ISBN: 978-0-429-46729-5 (ebk)

Typeset in Times New Roman
by Wearset Ltd, Boldon, Tyne and Wear

MIX
Paper from
responsible sources
FSC
www.fsc.org   FSC® C013056

Printed and bound in Great Britain by
TJ International Ltd, Padstow, Cornwall

In memory of Jean Gayon (1949–2018)

Contents

Tables

Acknowledgements

The philosophy of biology is haunted by a bearded figure: that of Mr. Charles Darwin, the solitary naturalist from Down. Assessing the measure of this man's unique achievements is a task that must be forever begun anew, and one that requires many collaborations and scholarly interactions.

In the long course of preparing this book, I have benefited from two grants, one from the French National Research Agency (ANR-07-JCJC-0073–0001) and one from the Institut universitaire de France (2011–2016).

I am most thankful to the advice and assistance of many colleagues and students, in alphabetical order: Richard Bellon, Richard Burian, Pietro Corsi, Richard Delisle, Jean-Marc Drouin, François Duchesneau, Jean Gayon, Jonathan Hodge, Lucie Laplane, Mathilde Lequin, Tim Lewens, Laurent Loison, Francesca Merlin, Karine Prévot, Gregory Radick, Marsha Rishmond, and Michael Ruse. Christopher Stevens has been a constant and reliable interlocutor in the long process of editing, rewriting and amending the book. Throughout, Andy Curran and the other members of the Valmont de Bomare Society have provided me with wise counselling and witty advice, especially in the final steps of revising the manuscript for publication.

The following book expands on the ideas of my *Darwin contre Darwin: comment lire* l'Origine des espèces? (Paris: Éditions du Seuil, 2009). I thank the Éditions du Seuil, and especially my editors, Jean-Marc Lévy-Leblond and Sophie Lhuillier, for their guidance and support. The general argument was modified in order to make more palatable to the Anglophone readership who were the various "Darwins" suggested by the original title. New chapters have been written in order to clarify the argument. I gratefully acknowledge the work of several anonymous reviewers whose comments had a great impact on the final version of the book.

The present work also relies on material that has previously appeared in various articles, which has been modified and integrated in the present version:

"The Evolution of the Origin (1859–1872)". In Michael Ruse (ed.), *The Cambridge Encyclopaedia of Darwin and Evolutionary Thought*, Cambridge University Press, 2013, pp. 158–164. Integrated as part of Chapter 1.

"Translating natural selection: true concept, but false term?" *Bionomina*, 3 (2011), pp. 1–23. Integrated as part of Chapter 2.

References to Darwin's *Origin* are cited as follows:

Origin, year of publication, page number, followed by a reference to Morse Peckham's Variorum edition: page number (# number of sentence).

For example: *Origin* 1859, p. 490, *Var* 759 (#270).

For other references (CCD, CDM, DNS, etc.), see the reference list.

Introduction

The meanings of "Darwinism"

When I first began to read the *Origin*, I was bewildered by the fact that Darwin seemed uninterested in finding natural selection actually at work in the present. As though he were entirely satisfied with seeing artificial selection at work in breeding, he called for no experimental confirmation of his theoretical principle of natural selection—almost as if the issue didn't concern him, or as if he believed the long summary given in the abstract of Chapter IV of his *Origin* was quite sufficient. Indeed, this absence of concern for any demonstration of natural selection at work in the wild should both surprise and intrigue any layperson drawn to reading the *Origin of Species*, whether this be out of mere curiosity or as a requirement on a college curriculum. It struck me, as a young philosopher, eager to find responses in Darwin's works to those who denied the existence of natural selection. But reading the *Origin* provided me with no straightforward rebuttal against those religious zealots and archaic Lamarckian biologists—there were still quite a few in France—who claimed that Darwin's *pièce de résistance* contained nothing more than mere hypothesis. And so I started reading more and more books and articles on Darwin and his work, to the point that I even translated the first edition of the *Origin* into French. In the process, I started to realise I did not know who "Darwin" was; I did not know what "Darwinism" meant.

But who is Darwin?

> Rudge immortalised the name of Darwin through his introduction to the Flora of the genus *Darwinia*.

This essentially anodyne sentence is adapted from an 1833 German publication, *Neues elegantes Conversations-Lexicon für Gebildete aus allen Ständen*. A nineteenth-century commentator nevertheless remarked that "[to] read these words nowadays is to be surprised by a variety of reflections".[1] Probably the first such surprising reflection is the idea that it may have been the English botanist Edward Rudge (1763–1846)—rather than Darwin himself!—who immortalised this now household name. Was the Darwin name really immortalised, not for the discovery of "natural selection", nor even for the description of a world in which species change, but instead because the name was given to a genus of Australian plants as far back as 1815?[2]

A second surprise follows swiftly as we begin to reflect on these dates themselves: the genus *Darwinia* was established in 1815, many years before the publication of the *Origin of Species*. This prompts the question: who is this "Darwin"? Today, of course, the name Darwin is employed to refer unambiguously to Charles Robert Darwin, born 1809, died 1882. But, as we shall see, well into the middle of the nineteenth century, "Darwin" referred to Charles' own grandfather Erasmus (1731–1802), renowned author of the *Loves of Plants*.[3]

In fact, the public may have encountered the words "Darwinian" and even "Darwinianism" published as early as 1804 in the *Edinburgh Review*, just two years after Erasmus had passed. The latter's style of poetry had gained fame among English poets, and the *Edinburgh Review* refers to "the originality of manner which is supposed to characterise the new Darwinian school of English poetry".[4] Indeed, that same year, the Lichfield poet Anna Seward (1742–1809) published a 430-page biographical "Memoirs of the life of Dr Darwin, chiefly during his residence at Lichfield". Seward described the old house Erasmus had bought in Lichfield as "a *rus in urbe* of Darwinian creation", surrounded by a philosophical circle of friends and peers, all forming a "Darwinian sphere".[5] Again in the same year, an English version of Virgil's *Georgics*, translated by William Sotheby, was accused of being run through with "Darwinian modulation" and "the false decorations which Dr Darwin has introduced in the common iambic measure".[6] This accusation of stylistic infection is pronounced using a quite picturesque phrase, perhaps coined especially for the occasion: a "charge of *Darwinianism*".[7] In private letters, reference to the "Darwinian style" of poetry was already in common use among English poets and literary persons from the 1790s on. Walter Scott, for instance, mentioned it in 1797.[8]

Dr Erasmus Darwin's name was also intimately associated with ideas about the transformation of species in the conversations of English high society circles. Benjamin Ward Richardson reminisced about such an evening as late as 1856, spent at Hartwell House, the home of astronomer and philanthropist John Lee, F.R.S. (1783–1866). Also present was the Rev. Joseph Bancroft Reade, F.R.S. F.R.M.S. (1801–1870), "one of the brightest men of science the Church of England ever produced"[9] and in no small way responsible for popularising the new photographical craze within this same circle. Captain Fitzroy was also a member of this group, together with Rev. Charles Lowndes, the chaplain of Hartwell, and the engineer and geologist Thomas Sopwith F.R.S (1803–1879). Nights were spent in the drawing-room, alternating between Mrs. Lee's excellent musical renditions and Dr. Lee's calls for somebody to give a short lecture or tell a story. On this particular evening, Richardson recalls, Sopwith

> described the hypothesis of the development of living things from a primordial centre. That, said [Rev.] Reade, is rank Darwinism. It was the first time I had heard that word used. It had no reference to Charles Darwin, whose name at that period was not connected with the subject; but it had reference

to Erasmus Darwin, and to his original and fruitful observations. I name this incident as indicating that Darwinism, like everything else, is itself an evolution.[10]

When Charles R. Darwin published his theories on the "origin of species", the name "Darwin" was already familiar currency within the scientific milieu, something which gave rise to many unconscious associations. In July 1837, Darwin himself opened his notebooks on the transmutation of species with an entry entitled "Zoonomia", an obvious reference to his grandfather, as the works of Jonathan Hodge and Phillip Sloan have made clear.[11]

It is thus evident that, in the mid-nineteenth century, the name "Darwin" already evoked a range of different things: a poetic style, fantastical reflections on the transformation of species, and, of course, the love of plants. While the fame of Erasmus Darwin certainly coloured the reception of Charles' works, the converse is also true, since the elder's writings enjoyed a return to fame once the works of his grandson had caught the public's attention.[12]

This introductory puzzle—playing on the similarity of names and plurality of Darwins—creates a doorway for us unto the perplexity of who "Darwin" was and what, then and now, "Darwin" really means. In other words, then as now, the name "Darwin" covered a plurality of meanings and references.

Readings: another "entangled bank"

Another striking fact requires that we "pluralize" Darwin: it is the effect of readings. Gillian Beer has taken up the question of culture in science, from the perspective of language.[13] Her work teaches us to perceive what Darwin's textual devices have in common with other nineteenth-century writers of fiction. Beer beautifully shows how Darwin's own language allowed his contemporary readers to assimilate the *Origin* according to various protocols. She emphasises how Darwin constantly rephrased his sentences, but also how "the Darwinian theory has ... an extraordinary hermeneutic potential—the power to yield a great number of significant and various meanings".[14] As Jeff Wallace put it:

> Darwin's language in the *Origin* is not so much his own as that of his culture: and again, not so much a unitary thing as a tapestry of discourses, borrowed or inherited, with varying degrees of mindfulness, from the evolutionism already evident in much social theory, from the natural theology the *Origin* is in other ways an answer to, from the literature, perhaps of Shakespeare, and certainly of Milton, Dickens and Wordsworth, as well as from a common stock of rhetoric which he raided in order to make the text the object of popular consumption he wanted it to be.[15]

But this constant process of rewriting and adapting to various audiences created a lot of ambiguities in Darwin's text and his readers were often keen to underline these difficulties, to the point that, like Daedalus' statues, "the object of popular consumption" would have escaped the control of its crafter.[16] Notoriously, in July of 1871, Darwin was presented with what would be one of the most ruthless attacks against his *Origin of Species*: St. George Mivart's *Genesis of Species*, the very title of which was an insolent stone cast at the greenhouse of Darwinism. After reading the book, Darwin complained bitterly about how he had been read and cited: Mivart quoted only beginnings of sentences torn away from their paragraphs, latched onto and extracted a handful of words from Darwin's "one long argument" and, in so doing, mutilated the actual meaning of the text. Immediately, Darwin sent a call to arms to the young American mathematician Chauncey Wright.[17] On 14 July 1871, Darwin wrote:

> I believe Mr. Mivart to be a thoroughly inaccurate man; but he was educated
> as a lawyer and seems to me to plead, as if retained by a client.—I detected
> in two places that he gives the commencement of a sentence or paragraph
> and by omitting the rest attacks my meaning. A review has just appeared in
> our *Quarterly*, evidently by Mivart, and cutting me into minced meat.

Darwin had the impression he was being turned into minced meat both metaphorically, since Mivart's principal aim was to destroy natural selection, and also somewhat literally, insofar as his sentences had been chopped into tiny bits and pieces.

Wright took up Darwin's challenge and set about responding to Mivart's objections. Darwin knew of Wright's research from as early as 1859 and Wright, in turn, had been a staunch supporter of Darwin ever since reading the first edition of the *Origin*. The young American's rebuttal to Mivart, published in the *North American Review*, impressed Darwin with its accuracy and rhetorical efficiency. In fact, he so admired Wright's paper he had it reissued at his own cost in Britain, as the rebuttal of reference to Mivart's *Genesis*. However, in reality, Wright did not respond to the actual *content* of Mivart's objections, such as, for instance, the puzzle of incipient structures. Rather, Wright plainly undertook a step-by-step analysis of Mivart's approach to both reading and citing.

In the *Origin of Species*, Darwin noted the scant number of domestic breeds of the goose. But instead of supposing that this bird had escaped the attention of artificial selection, Darwin commented on the feature specific to the goose: "the goose seems to have a singularly inflexible organisation".[18] For Mivart, Darwin's "attempts to explain this fact as regards the goose by the animal being valued only for food and feathers, and from no pleasure having been felt in it on other accounts" are left hanging, awaiting completion: in fact, Darwin seems very close to admitting "the conception of a normal specific constancy, but varying greatly and suddenly at intervals".[19] What catches Mivart's attention is Darwin's confession that a species may be endowed with "a singularly inflexible organisation": "Mr Darwin himself concedes the existence of an *internal* barrier to change". Mivart identifies in Darwin's texts hints of an "internal proclivity to change" as well as a symmetrical "marked proclivity to reversion"—both which seem to contradict the efficiency of natural selection in accounting for change. In

Wright's opinion, the goose structure is not *inherently* inflexible, but only *in comparison* to other animal structures and organisations.[20] Another area of mis-interpretation involved the use of polysemic terms. Mivart, for instance, inter-prets "plasticity" as a substantial force. But Wright strongly rebukes him: when Darwin refers to an organisation becoming "plastic", he means that it is suscepti-ble to being shaped and modelled.

But was Wright himself truly faithful to Darwin? An anonymous reviewer of Wright's rebuttal noted that the young mathematician placed too much focus on natural selection, whereas Darwin himself, in the *Descent of Man*, pronounced this important caveat: that he may, in previous editions of the *Origin*, have "attributed too much to the action of natural selection or survival of the fittest".[21] Starting with the fifth edition of his *Origin of Species*, Darwin stated his inten-tion to limit natural selection's sphere of action to "adaptive changes of structure".[22]

Since Darwin himself restricted the scope and effect of natural selection to just one among many causes of change, we are encouraged by the anonymous reviewer to always "[keep] in memory" that

> Mr Darwin himself, from the very nature of the process, has never supposed for it, as a cause, any other than a co-ordinate place among other causes of change, though he attributes it to a superintendent, directive, and controlling agency among them.

As a consequence, it may well be that "the place to which natural selection is entitled" could "be found less important than that assigned to it by Mr Darwin". Nevertheless, "the great service he has done on behalf of the general principle of evolution should not be forgotten".[23]

Mivart read Darwin and commented on his text. Wright read Mivart reading Darwin and then, in turn, commented on this reading. Next, an anonymous reviewer voiced his opinion on Wright reading Mivart reading Darwin, having himself also read some of Darwin's texts to conclude, quite paradoxically, that Darwin might actually be closer to Mivart than to Wright. Such entangled stacks of readings complexify the meaning of the word "Darwinism", the evaluation of Darwin's contribution to evolutionary theory, and the very existence of a so-called "Darwinian revolution".

Outline of the present book

My book follows a historical and philosophical approach. Without contesting the importance of natural selection to Darwinism, much less supposing that a fully-formed "Darwinism" stepped out of Darwin's head in 1859, its aim is to return to the text of the *Origin* itself, not to produce yet another orthodoxy, but in order to hold it up against the host of interpretations it has spawned. The present book is composed of three different parts. Each of these is preceded by a short intro-ductory section providing the reader with an outline of the main points. It starts with the necessity to go beyond the classical figure of "Darwin-the-Selectionist" drawing on three different sets of arguments: the plurality of versions of the *Origin*; the intrinsic ambiguities of Darwin's concepts, as revealed by the translations; the intricacies of Darwin's argument in the *Origin* (Part I). Then the book produces some alternative figures of Darwin as they were tracked down by readers in the body of the *Origin* itself: "Darwin-the-Variationist" and "Darwin-the-Lamarckian" (Part II). Finally, the figure of "Darwin-the-Cosmologist" is produced from other parts of the *Origin* (Part III).

Darwin-the-Selectionist and beyond (Part I)

Faced with a multiplicity of Darwins and Darwinians, it struck me as indispensable to get to the bottom of the conditions that had enabled such a plural state of affairs. To do this, and without in any way wishing to belittle the interest of the manuscripts and other texts, I resolved to re-immerse myself in the original *Origin*. Reaching a similar conclusion in 1979, James Moore decreed the following methodological principle: "a just understanding of Darwin's theory of evolution by natural selection must be founded on Darwin's statement of it".[24] Moore happily boasted that "fortunately" he had located "at least one passage in the *Origin of Species* that remained substantially unaltered throughout the book's six editions (1859–1872)", and which therefore could be deemed to "[distil] the essence of Darwinism into less than five hundred words": the summary placed at the close of Darwin's chapter on "natural selection"—a passage made striking by its overall hypothetico-deductive tone, with its repeated use of "if" and its eagerness to bring irrefutable facts to the table and then draw unavoidable conclusions from them; a passage also striking in its combination of two kinds of selection, both *natural*, leading to extinction, and *sexual*, leading to a competitive advantage in the mating process.[25] One may agree with Moore that readers of the *Origin* seeking out its core meaning should naturally turn to passages such as this since they offer, distilled, both the very spirit of Darwin's epistemology and the main thrust of his contribution to science. Readers who are first and foremost intent on finding pronunciations of Darwin's principal discovery are right to turn to those pages where the principle of natural selection is explained.

From Darwin's original publication in 1859 to us mere laymen striving to capture the spirit of Darwin's contribution to science today, the road is long and we may at times need guidance to find our way. French epistemologist Jean Gayon has provided us with a philosophical history of biology, showing how Darwinism emerged as a gradualist and panselectionist schema. This, Gayon claims, characterises "every interpretation of evolution as a gradual modification of species, predominantly orientated by a process of natural selection operating on the domain of intrapopulational variation".[26] Such a perspective is essential for understanding what has historically come to constitute "Darwinism" over the last 150 years since the original publication of the *Origin*.[27] What Jonathan Hodge has called "the traditional   evolution-plus-natural-selection-as-a-mechanism-for-it   formula   for

Darwin's thought"[28] constitutes a clear and efficient hypothesis on the nature of Darwin's text.

Jean Gayon starts his account on "Darwinian theories of evolution" with the following definition: "'Darwinian' refers to any interpretation of evolution that involves the gradual modification of species, predominantly guided by a process of natural selection operating on a field of intra-populational variation".[29] Enquiring into the history of ideas with this "Darwin" in mind, Gayon concludes on the fate of Darwin's ideas in their relation to the experimental status of the hypothesis of natural selection. If there were ever claims that Darwinism was dead, it is, Gayon perspicuously shows, because there was some "experimental refutation of the hypothesis of natural selection". What is natural selection? For Gayon (whom I follow closely on this point) "natural selection", as Darwin intended it, is a factor of progressive evolution, acting gradually upon infinitesimally small inherited variations, and constituting the dominant force or paramount power in the modification of species.[30]

Such "selectionist" readings of the *Origin* are supported by a sample of quotes, mostly from its first edition (Table I.1).

I should confine myself here to noting that the phrase "natural selection has been the main but not exclusive means of modification" (which Gayon quotes as evidence for the paramount power of natural selection) has been subject to quite contradictory readings, as the rest of this book will make clear.[31] While reading the *Origin*, one may indeed be struck by the fact that Darwin seems sometimes to undermine the centrality of natural selection. The philosopher Michael Ruse, in his own plain-speaking manner, has related a similar feeling of surprise, and

Table I.1 Quotes supporting Darwin-the-Selectionist

Progressive evolution	"This principle of preservation, I have called, for the sake of brevity, Natural Selection; and it leads to the improvement of each creature in relation to its organic and inorganic conditions of life" *Origin* IV, (b) 1860, p. 127, *Var* p. 271.
	Darwin later added: "and consequently, in most cases, to what must be regarded as an advance in organisation" *Origin* IV, (e) 1869, pp. 160–161, *Var* p. 271.
Acting gradually	"natural selection can act only by taking advantage of slight successive variations; she can never take a leap, but must advance by the shortest and slowest steps" *Origin* VI, (a) 1859, p. 194.
Infinitesimally small inherited variations	"Natural selection can act only by the preservation and accumulation of infinitesimally small inherited modifications, each profitable to the preserved being" *Origin* IV, (a) 1859, p. 95.
Paramount power in the modification of species	"Furthermore, I am convinced that Natural Selection has been the main but not exclusive means of modification" *Origin*, Introduction, (a) 1859, p. 6.

even confusion, at the way Darwin dealt with his concept of "natural selection" in the *Origin of Species*:

> At the time of writing the *Origin of Species*, Darwin had no direct evidence of selection. He strikes me as almost casual about the lack of such evidence.... Perhaps he thought, also, that the workings of selection would be so slow it would be difficult to capture it in action.[32]

Perhaps Darwin also had in mind that he had to solve some issues that were more important than natural selection before he could clearly demonstrate the importance of his mechanism. Ruse suspects that Darwin disposed with an experimental confirmation of natural selection because he was fully assured of the methodological and epistemological status of his natural selection hypothesis—we shall return to this question of natural selection as a "*vera causa*". But in my own reading of the *Origin* I was struck by quite a different fact. Darwin did call upon his readers to engage themselves in further study and experimental field work, only he called for their endeavours to focus not on the empirical status of natural selection but on something else entirely: *the causes and laws of variations.*[33] *Variation*, not Selection, is the field he laid open for his followers to explore. From this realisation, a new question began to grow in my mind: what if natural selection were not the only principle underlying Darwin's work *On the Origin of Species*?

Darwin's writings are complex and not reducible to the picture of him presented in extreme selectionist (or "Neo-Darwinian") versions of his theory—such as Wallace's and Weismann's, or that espoused by the Modern Synthesis—whereby natural selection is the sole legitimate explanatory cause within evolution. My goal has instead been to "de-Synthesise" Darwin: to approach Darwin without being guided by the standard view the theoreticians of modern population/evolutionary biology—the founders of the Modern Synthesis—have acclimatised us to. It goes without saying that natural selection is the mainstay of Darwin's work; we not only associate his name with it, but no other concept better captures what is generally meant when we speak of "Darwinism". And yet, there is much more to the story than this. The case this book argues is that while Darwin did of course propose a theory of natural selection, his *Origin of Species* can just as accurately be seen as a contribution to the study of variation and as the grounding for a new cosmology. Darwin-the-Selectionist is not the sole inhabitant of the *Origin*. Among the plurality of Darwins the *Origin* contains, one is of particular interest to us: Darwin the theoretician of variation, the ponderer of the conditions that bring it about, the seeker after the laws that govern its manifestation. My wager? Scratch the selectionist surface and watch the variationist bleed. Variation is not a handmaid to natural selection, merely providing the raw material for selection to act upon, as the standard view of the Darwinian mechanism would have it. Rather, it is a field of research unto itself.

Darwin-the-Variationist (Part II)

What effects does this hypothesis produce when reading the *Origin*? This hypothesis brings to light the fact that Darwin develops Chapter V, dealing with the laws of variation (a chapter often left aside in reconstructions of the book's argumentation that focalise on the analogy between natural and artificial selection), in direct continuity with the celebrated Chapters III ("Struggle For Existence") and IV ("Natural Selection"). If the *Origin* does contain a theory of variation, then Chapter V would have to be returned to its pivotal position in the economy of the book. Moreover, variation also places Darwin himself in continuity with a concern for the laws of organic life inherited from old grandpa Erasmus.

Suspending the hypothesis of natural selection and the various reformulations it has undergone over time, I intend to take you along what may be considered a dark and potentially bewildering path: heading straight into the full complexity of the *Origin*'s text, endeavouring to catalogue threads other than just selection along the way. It is not a mission we set out on lightly, with neither reason nor guide, since, in reality, it was as a contribution to the theory of variation that the *Origin* was read and understood when first published.[34] The point was decisive for one very simple reason: for Darwin's contemporaries, in order for natural selection to take place, there needed to be plentiful stock of "undetermined" (the significance of this term will be examined further on) variations.

One commonplace suspicion that may be present in some readers' minds must also be addressed: the sneaking thought that if Darwin was not merely a "Darwinian", then this must surely mean that this *other* Darwin must have been a Lamarckian. But since this is not the conclusion my research led to, I must dispel this illusion.

First, Darwin's relation to Lamarck is complex and he argued vehemently against any amalgamation of their respective theories. Darwin writes Lamarck into the "History of Error", and to those who like to think the *Origin* offers nothing more than "Lamarck's views improved by [Darwin's]",[35] the latter retorts that the author of *Philosophie Zoologique* made no original discoveries, noticing and relating only the obvious; general theories even Plato had been able to formulate. For this, Lamarck's volume is written off as "a wretched book", "one from which (I well remember my surprise) I gained nothing".[36]

Besides, the word "Lamarckism" doesn't help us here. The intellectual movement known as Lamarckism can be of course identified with the ideas that Jean-Baptiste Monnet de Lamarck published in his *Philosophie Zoologique* (1809) fifty years before the *Origin*; but the history of Lamarckism is a complex one. Pietro Corsi has demonstrated what Lamarck's fame owed, especially in Britain, to the second volume of Charles Lyell's *Principles of Geology*.[37] Peter Bowler even claimed that Lamarckism in France might well stem from the introduction of post-Darwinian German ideas—specifically Ernst Haeckel's.[38] To many thinkers, Henri Bergson (*Creative Evolution*, 1907) among them, Lamarckism is an American tradition, through the works of Edward Drinker Cope, Alpheus Hyatt, or Alpheus Packard, who apparently coined the term "Neo-Lamarckism".

In its American version, Lamarckism embraces mystical theories of psychic forces driving evolution (bathmism, mnemogenesis).[39]

Just like "Darwinism", the theoretical entity known as "Lamarckism" has fuzzy boundaries. It acts more like a catch-all term for any evolutionary idea that does not accept natural selection as its central mechanism.[40] "Lamarckism" can designate the theory which states that evolution's driving force resides within *the will* of individuals. It can indicate a theory of soft inheritance, where use and disuse, or environmental conditions, play a triggering role in the transformation of living forms. Recent studies by Laurent Loison of French neo-Lamarckians (like Alfred Giard, Gaston Bonnier, Julien Costantin, or Maurice Le Dantec) have revealed them to be heirs of Louis Pasteur and Claude Bernard: they experimentally studied different ways in which the environment affects the shape and development of biological individuals.[41]

As a result, I intentionally avoided referring to "Lamarckism", a hodgepodge of various non-selectionist theses—few of which could in fact be interpreted as being *Lamarck's* own theses. Indeed, as far as possible, I have tried to avoid any such general terms as "Darwinism" or "Lamarckism" as I believe them to be a posteriori reconstructions which serve only to obfuscate real debate by posting easy-to-identify labels on the theoretical positions at play.

Given this fuzziness and all-encompassing character of the "Lamarckian" theoretical entity, it would of course be easy to isolate "Lamarckian" elements in Darwin's works. As I have already mentioned, this would be true even of the first edition of the *Origin* (1859), and it has been claimed that the case for a "Lamarckian Darwin" only gets stronger with each subsequent edition, reaching a pinnacle in the final 1872 text.[42] Darwin undoubtedly included more than his fair share of "Lamarckian" elements in his writings: Chapter V on the "laws of variation" is the *locus classicus* one would inhabit in order to be fully surrounded by Lamarckian-esque notions on the effects of use and disuse in inheritance, etc. However, one must never forget that Darwin constantly distanced himself from Lamarck's "wretched book" and its imperfect ideas. For all these reasons, it would be a violence to history and quite simply incorrect to equate Darwin with Lamarck. The case of a progressive "Lamarckianisation of Darwin", increasing with each successive edition of the *Origin*, is often alluded to. But I doubt this conclusion as it leans on fragile evidence: some of Darwin's additions to the *Origin* emphasised the role of natural selection while others stressed the importance of still other means of modification. In conclusion, a "Lamarckian Darwin" would be a very complex figure, one that would most likely create more problems than it solves. So then, there were not only two Darwins—namely, a "Darwinian" one and a "Lamarckian" one—but more.

Darwin-the-Cosmologist (Part III)

"Darwinism" is often understood as the name of a new philosophy: it questions the limits of knowledge, causality or teleology. Reflecting this, many readers contemporary to Darwin called for his theory not to be criticised for its methodological foundations (it is meaningless to relegate it to the status of mere "hypothesis" since hypotheses are a normal part of how science works) and instead celebrated it as a new metaphysics. These same voices appealed for his work to be analysed according to the coherency of the system it proposed.[43]

Beyond the interest for transformism and evolution in general, what seems to characterise all contemporary reactions of the time is an eagerness to seek laws. If we judge, for example, by the French titles given to the first translations—*De l'origine des espèces, ou Des lois du progrès* [laws of progress] *chez les êtres organisés* (tr. 1862); *De l'origine des espèces par sélection naturelle, ou Des lois de transformation* [laws of transformation] *des êtres organisés* (tr. 1866)— in each occurrence, the "law" element is introduced into the title as if the only matter of genuine importance was that Darwin had discovered new laws of nature. This goes hand in hand with a double criticism belonging to readers of Darwin who considered the *Origin* to err in two ways: first, on the specific point of the origin of species, the mechanism Darwin proposes is unsatisfactory and requires completion; but, second, and more importantly, Darwin's question of the origin of species appears to be wrongly framed, since broaching this topic depends on broaching other, more fundamental questions, such as the origin of variations, the origin of life, or the origin of mankind. In both cases, this leads to a search for new laws, among which the establishment of the laws of variation the *Origin* dedicates its fifth chapter to, yet nonetheless leaves incomplete.

As a matter of fact, Darwin's *Origin* doesn't have an answer for everything. It doesn't worry itself to define species at the beginning of Chapter II.[44] Darwin wilfully distances certain points and opens his chapter on instinct with the declaration, "I must premise, that I have nothing to do with the origin of the primary mental powers, any more than I have with that of life itself".[45] But all of these precautions were in vain. Indeed, it is as if, even before Darwin's work was published, a certain number of questions had been tied, as though necessarily, to any theory of the origin of species by natural transmutation instead of by special creations that might come forth. Among these questions, two come regularly to the fore: the origin of life, and the origins of mankind.

Revisiting the *Origin of Species*

Another question has to be asked before writing on Darwin: why add to the torrent of Darwinian literature at all? As Michael Ruse put it: "Darwin Studies is a field very well plowed indeed".[46] Almost forty years ago (1979), the British historian James Moore opened his book with words that still ring an all-too-familiar bell with the historian or philosopher undertaking to write a book on "Darwin" and his *Origin*: "In the present study we endeavour to penetrate the bibliographic jungle surrounding the post-Darwinian controversies." "Now", Moore adds, "it would not only be presumptuous but plainly self-defeating to disparage the existing literature of the post-Darwinian controversies".[47] My book also draws on this immense and fascinating literature to provide its readers, in a limited number of pages, with a map of the many ways in which the *Origin of Species* has been read.

The boundless erudition of the "Darwin Industry" has trained us into thinking that any thorough and sound work on Darwin should follow him step by step from the Beagle voyage, all along the winding ways traced in his notebooks, via the comparison of different versions of his theory, and, stopping off at each of the multiple editions of his works and their numerous variations en route, should then continue right up to the great man's death. "From the Beagle to the Grave" would have been a beautiful title, had I followed that route, closely following in the footsteps set down by Janet Browne in her masterful two-volume biography.[48]

However, in spite of the enormous bulk of Darwinian literature, many lay readers deciding to read Darwin's published work do not know every detail of Darwin's thoughts and nonetheless start reading his book. A great majority of readers open the *Origin* with little or no prior knowledge of what Darwin's thought was in 1837, 1844, or 1872. They are simply curious to read one of the milestones of scientific literature. As a result, and despite Darwinian scholars publishing thousands of pages demonstrating to us that Darwin's world was nothing like our own, generation after generation, students of biology and philosophy, laymen, curious minds of all types still open up a popular edition of the *Origin*, such as the Penguin paperback, expecting to read the thoughts of a contemporary, someone they will understand at first reading, the familiar silhouette of the bearded old man on the cover. More importantly, they expect the book to contain a fully developed theory of evolution. Against this reality, an annotated academic edition, like that edited by James T. Costa for Harvard University Press in 2009, occupies an ambiguous position: its footnotes aim to place Darwin in relation with recent discoveries in biology while other more erudite notes grant the reader glimpses into old-fashioned theories unknown to anyone today, barring a handful of historians of science.[49]

This is why I chose to re-immerse myself in the *Origin*. But which edition of the *Origin* was I to choose? Here again I encountered the great figure of Ernst Mayr and the modern evolutionary synthesis, and their defining role in Darwinian debates. Above all, when it comes to the *Origin* itself, today's scholarly consensus opts for the first edition from 1859. This preference has been well recognised ever since Harvard University Press republished the first edition of the *Origin* in facsimile in 1964, complete with a preface by Ernst Mayr. In Mayr's view, the reason we should return to the initial eruption of this artefact onto the scientific scene, is that "the publication of the *Origin of Species* ushered a new era in our thinking about the nature of man", thereby sparking an intellectual revolution, "greater than those caused by the works of Copernicus, Newton, and the great physicists of more recent times". It is in the first edition that we find Darwin's thought in its pure state, packing its full iconoclastic punch. As Mayr puts it, "Chastened, Darwin softened his statements and withdrew some claims in later editions". Such attenuation ultimately rendered the sixth edition unfit for penetrating the meaning of Darwin's true thoughts.

Over the course of the various editions, Darwin made all sorts of modifications, from simple comma corrections to considerable reorganisations. What's more is that he attached great importance to these changes, even though today they are often seen as mere parasitic appendages. Present-day readings of the *Origin* set their sights on a "pure" Darwin. This goal preys on the minds of more than a few readers, and for this ill there is but one available tonic: to return to the original text.[50] Lurking behind these concerns with establishing a definitive version, the root cause of disquietude stemming from later editions of the *Origin* is a preoccupation with distinguishing, as much as humanly possible, "Darwinism" from all forms of non-Darwinian evolution, such as the "orthogenetic" doctrines which posit "laws of necessary development" at work in evolution.[51]

But in fact, even Mayr, in his optimistic comeback to the first edition, had to come to the conclusion that the original 1859 edition itself was enough to produce perplexity in the reader. The pristine 1859 edition turned out not to be the ultimate manifest of Darwinism, as Darwin himself appeared not to be the unconditional and orthodox defender of natural selection he was expected to be. From the time of its first publication in 1859, *On the Origin of Species*, Darwin's masterpiece, already contained its fair share of ambiguities and diverse aspects, each of which has been extensively underlined, commented, and criticised. While the *Origin* did rapidly establish itself as an indispensable book and, more broadly, as one of the most remarkable works of Western science, it also gave immediate rise to innumerable different readings, in turn multiplied by the successive editions and translations, each and every one of which poses its own problems. What does it mean to be a Darwinian when we know that the role of natural selection was to some degree attenuated by Darwin himself, beginning with the 1859 version and reinforced with ever more insistence thereafter?

Re-immersing myself in the *Origin*, my intention was neither to assemble the history of Darwin's ideas nor to try to understand how his theory came together and why it remains so important to us today—admittedly two fascinating and often explored questions. Rather, I did so in order to solve a puzzle: how is it that a single text, the *Origin of Species*, can lend itself to so many contradictory interpretations? In the present book, I have tried to use historical material to change our outlook on the *Origin of Species*. I didn't want to write a book that frames the *Origin* in its relation to the biology of the present day. Instead, I wanted to write a book that restores the *Origin of Species* to its status as an inextinguishable source of puzzlement and bewilderment for its readers.

To accomplish this goal, I have to make a series of important claims in order to describe the framework that directed my reading.

First, the *Origin of Species* does not stem from one, unique theory or argument. Of course, the book decisively presents all manner of facts and arguments supporting the view of "evolution", or "descent with modification" by means of "natural selection", which is to say it presents a stance contradicting the view of "creation" (terms I will return to later). But the *Origin* cannot be reduced to "one" argument. Ultimately, this has come to be widely accepted in Darwinian studies, ever since Ernst Mayr famously spoke of "Darwin's five theories".[52] It is perfectly fine to say that natural selection has achieved scientific status; the invention of this concept explains why Darwin's contribution to science has stood out for more than 150 years. But Darwin cannot be cast as an ultra-selectionist. He believed in the efficiency of many other causal mechanisms, and his *Origin* is full of notations that support this case.

Second, "Darwinism" is an unclear, historically fluctuating set of theories or hypotheses. This has been studied by, among others, David Hull, James Moore or David Depew and Bruce Weber.[53] An atheist or materialist Darwinism can be defended, just as there is a coherent Christian Darwinism, or, if one prefers, a "Christian Darwinisticism"—a neologism coined by Morse Peckham to suggest a fusion of Darwinism and romanticism.[54] There have been gradualist Darwinians but also saltationist disciples too. In short, any narrow definition of Darwinism will always be incorrect for there are always plenty of exceptions to be found. Puzzled by the term "Darwinism", Jacques Roger once bluntly asked: "I wonder whether it would not be simpler to start with Darwin himself and what he wrote than with an imaginary Darwinism whose existence in the realm of ideas creates unnecessary difficulties?"[55] However, it is hard to avoid speaking about "Darwinism", and, indeed, everyone does refer to this term. Friends and foes of evolutionary ideas all refer to "Darwinism". August Weismann, Alfred Russel Wallace, Clémence Royer, and many others all devoted papers, books, and encyclopaedia entries to "Darwinism". But we suspect that their "Darwinisms" are always attempted *coups* or take-over bids: they claim to redefine whole disciplinary fields under their own views. It may well be that "Darwinism" is nothing more than a straw man that no one actually believes in. As Jean Gayon put it: "any historian intending to deal with all the historical manifestations of "Darwinism" has first to tackle their extreme heterogeneity".[56] As Gayon's book masterfully shows, we should not buy into the epistemological fable of a "Darwinism" that developed smoothly from 1859 onwards. Darwinism is a contested field involving various attempts at a new "synthesis" as well as proclamations of the "death" of Darwinism in favour of some emergent general theory of evolution. The title of Eberhardt Dennert's 1903 pamphlet, *At the Deathbed of Darwinism*, was a successful motto,[57] although as Vernon L. Kellogg perspicuously noted in 1907: "ever since there has been Darwinism there have been occasional deathbeds of Darwinism on title pages of pamphlets, addresses and sermons".[58] Peter Bowler, in his *Eclipse of Darwinism*, has shown that this is an enduring tradition, since present day opponents of Darwinism still adopt the same posture: claiming that they are struggling against an oppressive "Darwinian" dogma, claiming to bring other (Lamarckian?) factors back into the evolutionary framework.[59] Kellogg, Bowler, and Gayon's analyses have shown us that there is no room for the enshrinement of a Darwinian dogma that exists outside of time and space. All reference to Darwinism must be apprehended as a potential source of controversy, sometimes rolled out as a slogan, other times as a foil serving the ends of other positions. For these reasons, we would be wise to adopt David Hull's stance when he concludes that "Darwinism" is a historical entity that must always be apprehended in a given context in time and space.[60]

Third, the history of post-Darwinian times cannot be reduced to the development of the theory of natural selection. In Darwin's time, several dimensions, already mentioned above (variation, inheritance, natural selection), were sources of puzzlement and inquiry. So, the history of post-Darwinian biology cannot be equated with the history of natural selection. While I can see why Jean Gayon writes that, after Darwin published his *Origin*, "the first task was to find empirical evidence for natural selection",[61] my feeling is that there were other major problems facing biologists. Clearly, this is not what Darwin writes in the *Origin*, and this is not what most of his readers and followers understood that they were expected to do. Since the history of natural selection has already been written, the present study focuses on other problems like variation.

Fourth, "Darwin" should not be considered as a man but rather as a text. We now know so much about the man that our interpretations of what he wrote and published have become heavily constrained as a result. Of this, I am absolutely aware. But I want to read Darwin-the-text *as if* we knew nothing of his inner qualms. In many ways, my book can be read in the same way as Albert Schweitzer's *Quest for the Historical Jesus*.[62] At first sight, the Darwin and Jesus cases cannot be more different: we know (almost) nothing of the historical Jesus, we know (almost) everything of the historical Darwin. Where we have not a single word written by Jesus himself, we have stockpiles of notes in Darwin's hand, allowing us to indulge in a full day-by-day reconstruction of his emotions and thoughts. Darwin-the-man is fully accessible to us. However, there are many parallels that could be drawn between Jesus scholarship and its Darwinian counterpart—not least of which, the enormous bulk of work that has been devoted to both "great men". A key issue in Jesus scholarship is the question of priority and authenticity between the gospels, especially Matthew and Mark's gospels, which are considered to contain the oldest evidence. In Darwin scholarship, we have no problem in minutely dating the various documents that are kept in Cambridge University Library; however, a comparable problem arises from the fact that we don't know how to weigh the *quality* of these documents, to prioritise or create a hierarchy between various versions, whether composed at different or more or less the same moments of Darwin's life, according to whether he is writing in different passages of his book, or in letters to various correspondents, etc. As previously noted, the formidable "Origin of the *Origin*" question emerged from the substantial body of sources made up of Darwin's autobiography, his notebooks, his day-by-day correspondence, the different manuscript versions of his theory, and the many notes dotted about the margins of his books, journals, and offprints. This constitutes both a colossal sum of documentary material and an enormous research archive into which the scholars of the last decades have thrown themselves. Has this enriched our well of information? Without a shadow of a doubt. And our assuredness? Not at all. The autobiography, for instance, is taken by some to be the final confession, the text that provides all the keys. For others, it is a hindsight reconstruction that exaggerates certain details and omits others. The same could be said for each and every letter of his correspondence. For example, when Darwin writes to his editor John Murray, telling him that his book on orchids will embellish on and "do good to the *Origin*" but will also be "like a *Bridgewater treatise*" (an example of a certain type of treatise in natural theology that was characterised by their intention to demonstrate God's infinite wisdom through knowledge of nature), just how are two such remarks to be reconciled? Is Darwin sincere in daubing his own work with the "Bridgewater treatise" brush? Or is this allusion just the promise of a bestseller, intended only to titillate his editor? In the face of such a multiplication of perspectives from within Darwin's manuscripts and letters, the text of the *Origin* itself offers scant reassurance: the innumerable differences between the six successive versions of the published text only add legitimacy to each of their respective claims to being the *proper* edition of reference.

With every successive biographical stage, a new Darwin steps forward: there is the Darwin of 1837 (the writing of his first notebooks), of 1838 (reading Malthus), of 1842 (writing the *Sketch*), of 1844 (writing the *Essay*), of 1854 (starting his *Big Species Book*), of 1859 (first publication of the *Origin*), of 1872 (sixth edition of the *Origin*), plus many others besides—the Darwin of the unpublished manuscripts, for instance, the Darwin of the margin notes, or the Darwin of the correspondence—a whole host of Darwins whose mutual coherence we can no longer be sure of. Is one the genuine article while all the others are false Darwins? While Dov Ospovat focused on Darwin's London years (the 1830s), Peter Vorzimmer focused on the post-*Origin* years: as a result, their views on Darwin are totally different, but can one then conclude that one is right and the other wrong?

On top of this, we also have to deal with a multitude of readings of Darwin, each one latching on to some single sentence and then marching under its banner. "Most readings", James A. Secord wrote, "leave little or no trace—an ownership signature or a few pencil marks. Only certain types of reading, such as academic study and reviewing, produce more substantial records".[63] These academic readings and reviews form the raw material from which I draw a sceptical or perspectivist lesson. Reading is a process of "defamiliarisation", as Jeff Wallace has put it: "while helping us to see the *Origin* anew, [such defamiliarisations] also reveal that there is no thing itself to which defamiliarisation can give us access—no 'stone' to be seen in its original 'stoniness' "[64] Readings are multiple points of view diffracted from a unique object, in itself unattainable.

All things considered, it seemed to me that the task required consisted of returning to the *Origin*, armed now with all that the manuscripts and conceptual situating of Darwin (framed by his predecessors and successors) have taught us. Returning from the tributaries to the source of the work, settling back into the base camp after the long Darwin scholarship expedition, this is the task I set myself, with Huxley's insightful remark from 1864 ever-present in my mind: "It is singular how differently one and the same book will impress different minds".[65]

My book provides a platform for precisely this apparent paradox: one does not need to be "anti-Darwinian" in order to take on the "Darwinians". Often, the historian who decides to document the conflict between Darwinians and their enemies (those other "evolutionisms", presumed to be responsible for the "eclipse of Darwinism") is led to observe that, in reality, it was actually opposing appropriations of Darwin that were battling it out. Certainly, none of these appropriations would fit with our current conception of what Darwinism is, but neither are any of them overtly anti-Darwinian. Most often, scholars in the second half of the nineteenth century were united in their praise for Darwin's contribution and saw their own efforts as a continuation of his work. Was this allegiance to Darwin just a "toll" that had to be paid? Mere lip service to be offered up in order to gain safe entry into the inner circle of Victorian naturalists? Perhaps, although not always. Even the most fervent Darwinians are not without heterodoxy, while the most vehement anti-Darwinians accord some impact (albeit a local one) to natural selection, and that those individuals most suspected of Lamarckism can surprise by their unconditional Darwinian allegiances (G.J. Romanes being the most singular example of this).

The other Darwins

Apart from the Selectionist view on the *Origin of Species* and its legacy, other readings of Darwin's major work have been and still are possible. This challenging book instantly found a wide audience of readers.[66] And yet its popular reception largely dispensed with natural selection, something Alvar Ellegård's 1958 study of popular journals has ably demonstrated.[67] But it appears that the scientific reception of Darwinism also regularly dispensed with this key aspect. It is as though Darwin's work was simply absorbed into debates on evolution that had begun long before it arrived.[68]

Indeed, "Darwinism" must not be seen as some intangible theory, nor the *Origin of Species* as a monolith that delivered the "truth" of modern biology. Rather than renewing praise of the *Origin* as a monument, it seems more useful to approach it as a complex philosophical object, subject to rival interpretations, and to thoroughly explore its every twist and turn. I do not intend to form another new dogma of Darwinism. Instead, I try to portray a surprising and paradoxical picture where it is Darwin himself who emerges as the source of inspiration to all those who have contested the significance of his central concepts or who have claimed to surpass him. I aim to show the complexity of the historical situation from which Darwinism arose, a process during which even Darwin swung back and forth between positions, opening doors through which several rival traditions were then able to charge. With this aim in mind, I have set out to retrace the fate of a legendary book, to describe the multiple lives it took on through its many editions, translations, and interpretations, right up to the captivating power it still exercises today. I strove to identify the various niches where rival interpretations have taken root; to unearth the ambiguities readers of Darwin latched onto as they produced their myriad Darwinian legacies, each more or less faithful to the originator's thought, each ultimately as monstrous as the other.

Notes

1 Stirling 1894, p. 1.
2 Rudge 1815.

3 On Erasmus Darwin, see for instance, Ruse 1999, pp. 37–53.
4 *Edinburgh Review*, April 1804, pp. 230–241, here p. 238.
5 *Ibid.*, p. 234.
6 *Ibid.*, p. 297.
7 *Ibid.*, p. 297.
8 Walter Scott to William Taylor, 22 January 1797, in Grierson 1932, vol. 1, p. 62. See also, for instance, Elizabeth Barrett Browning, *The Book of Poets* (1842) which refers to "a broad gulf between his (Wordsworth's) descriptive poetry and that of the Darwinian painter-poet school" (New York, J. Miller, 1877, p. 106). For other occurrences, see Oxford English Dictionary, entries Darwinian, Darwinianism and Darwinism.
9 Richardson 1891, pp. 254–256.
10 *Ibid.*, p. 256.
11 See Hodge 1985, p. 238; Sloan 2003.
12 In 1879, a translation of an *Essay on the Scientific Works of Erasmus Darwin* by Ernst Krause was published, now including an appendix by the grandson Charles: Erasmus Darwin, he remarks, had an "overpowering tendency to theorise and generalise" (Darwin 1879, p. 48). Samuel Butler (1879) also referred to the grandfather's publications.
13 Beer (1983) 2000.
14 *Ibid.*, p. 8.
15 Wallace 1995, pp. 12–13.
16 It is said that the statues crafted by Daedalus looked so much infused with life that they would have escaped if they had not been carefully bound to the wall. See Plato's *Meno*, 97d.
17 On Wright, see Madden 1963.
18 *Origin* 1861, p. 43, *Var* 116 (#310:c).
19 Mivart 1871, p. 119.
20 In Wright's terms, Mivart had succumbed to what the old scholastics called *fallacia a dicto secundum quid ad dictum simpliciter*—the mistake of stating in an absolute fashion (*simpliciter*) that which is only relatively true (*secundum quid*).
21 Darwin 1871, vol. 1, p. 152.
22 *Ibid.*
23 Anon. 1872, p. 262.
24 Moore 1979, p. 125.
25 *Origin* 1859, pp. 126–128, *Var* 270–272.
26 Gayon (1992) 1998, p. 1.
27 See Gayon 2009a and 2009b.
28 Hodge 1985.
29 Gayon 1998, p. 1.
30 *Ibid.*, p. 2.
31 See especially Chapter 6, below.
32 Ruse 2008, p. 75.
33 See *Origin* 1859, p. 486.
34 This crucial dimension of Darwin's work has been barely emphasised, especially in contrast to the huge amount of literature devoted to natural selection, or even to the sister topic of heredity. Before my own recent attempt emphasising "Darwin's failed Newtonian program" (Hoquet 2014), the question of variation is the object of two pioneering papers by Peter Bowler (1974 on Darwin, 1976 on Wallace) and by Rasmus G. Winther (2000 on Darwin, 2001 on Weismann). Olby (2009) bears less on variation as such than on the well-worn "What-if-Darwin-had-know-about-Mendel" question. Hallgrímsson and Hall 2005 bears on variation as a "central concept in biology" and includes an historical overview by Bowler.
35 Charles Lyell to Charles Darwin, 11 March 1863, CCD 11 218.

36 Cf. Darwin to Huxley, 9 January 1860, CCD 8 26; Lyell to Darwin, 11 March 1863, CCD 11 218; Darwin to Lyell, 12 March 1863, CCD 11 222–223.

37 Corsi 1978.

38 Bowler 1988, pp. 67–69.

39 Pfeifer 1965.

40 Bowler 1988, pp. 58–106. For a recent use of the term "Lamarckian", see for instance Jablonka and Lamb 1995, and for an overview on Lamarckism, see Gissis and Jablonka 2011.

41 See Loison 2010; 2012.

42 Liepman 1981.

43 For example, Wigand 1874, vol. 1, p. VIII and 9.

44 *Origin* 1859, p. 44.

45 *Origin* 1859, p. 207, *Var* 380 (#5).

46 Ruse 2013, p. xv.

47 Moore 1979, p. 6.

48 Browne 1995 and 2003.

49 For example, Edward Forbes, "the man who summoned up lost supercontinents to explain the distribution of species", as Desmond and Moore put it (1991, caption to picture 49).

50 All six editions of the *Origin* are now available on the site darwin-online.org.

51 For a complete presentation of non-Darwinian theories of evolution, see especially Bowler 1983 and 1988.

52 Mayr 1985.

53 Hull 1985; Moore 1991; Depew and Weber 1995.

54 Peckham, "Darwinism and Darwinisticism", in Peckham (1959) 1970.

55 Roger 1976, pp. 483–484.

56 Gayon 1998, p. 4.

57 Dennert 1904.

58 Kellogg 1907, p. 1.

59 See for instance Jablonka and Lamb 1995.

60 Hull 1985.

61 Gayon 1998, p. 11.

62 Schweitzer 1954. This enlightening parallel was suggested to me by an anonymous reviewer of the manuscript.

63 Secord 2000, p. 336.

64 Wallace 1995, p. 6.

65 Huxley 1893, vol. 2.

66 See Topham 2004 on the role of Charles Edward Mudie's ambulating library.

67 Ellegård 1958.

68 See, notably, Corsi 1988b; Secord 2000.

Part I

Darwin-the-Selectionist and beyond

1   A labyrinthine *Origin*

A first source for the wide variety of interpretations Darwin has been submitted to is the sheer number of editions of his seminal book, *On the Origin of Species by Means of Natural Selection*, that he oversaw.[1] Historically, depending on whether readers were referring to one edition or another, they could easily play one Darwin against another. A clear example of this underlies a series of debates that erupted in Christchurch, New Zealand after Samuel Butler published an anonymous review of the *Origin* in the columns of the newspaper *The Press*. An anonymous opponent, later known as "The Savoyard" (probably Dr Abraham, the Bishop of Wellington), concluded his subsequent critique of Darwin's book with the following remark:

> All his fantasias … are made to come round at last to religious questions, with which really and truly they have nothing to do, but were it not for their supposed effect upon religion, no one would waste his time in reading about the possibility of Polar bears swimming about and catching flies so long that they at last get the fins they wish for.[2]

Four days later, another anonymous critic, calling himself "A.M." (perhaps Samuel Butler himself), came back with this retort to

> the implicit statement that Darwin supposes the Polar bear to swim about catching flies for so long a period that at last it gets the fins it wishes for […] Now, however sceptical I may yet feel about the truth of all Darwin's theory, I cannot sit quietly and see him misrepresented in such a scandalously slovenly manner. What Darwin does say is that sometimes diversified and changed habits may be observed in individuals of the same species; that is, that there are eccentric animals just as there are eccentric men. He adduces a few instances and winds up saying that "in North America the black bear was seen by Hearne swimming for hours with widely open mouth, thus catching—almost like a whale—insects in the water". This and nothing more.

A.M. then adds a page number to his argument, referring readers to the *Origin*: "pp. 201 and 202". "The Savoyard" duly reacted to these accusations of "carelessness" and his being "disgraceful" on 11 April, and he too quoted from Darwin:

> Even in so extreme a case as this, if the supply of insects were constant, and if better adapted competitors did not already exist in the country, I can see no difficulty in a race of bears being rendered, by natural selection, more and more aquatic in their structure and habits, with larger and larger mouths, till a creature was produced as monstrous as a whale.

The editor of *The Press* then got involved, adding this perplexed note:

> The paragraph in question has been the occasion of much discussion. The only edition in our hands is the third, seventh thousand, which contains the paragraph as quoted by "A.M.". We have heard that it is different in earlier editions but have not been able to find one. The difference between "A.M." and "The Savoyard" is clearly one of different editions. Darwin appears to have been ashamed of the inconsequent inference suggested, and to have withdrawn it.

This was indeed the case: the page numbers indicated by A.M. refer to the third edition (pp. 201–202), whereas the paragraph cited by The Savoyard came from the first edition (p. 184). In this episode involving the bear/whale story, and on many other occasions like it, it is the various versions of the text that come into conflict, opening the way for vastly different readings and competing visions of what Darwin *really* had in mind. Darwin chose to delete a speculation that proved to be embarrassing, but not before critical damage had been done, as Alvar Ellegård has also pointed out.[3]

Why did the *Origin* change so much? Darwin had been elaborating his theory since 1837 and had been conscientiously working, since at least 1854, on what has been called his "big species book written from 1856 to 1858",[4] a book he referred to under various names, expressing his desire to produce a "Species theory"[5] or "to write a book with all the facts and arguments, which I can collect, *for and versus* the immutability of species".[6] The project was also given different code-names, from the quite vague "my big book"[7] to the last and most famous of them: "Natural selection".[8]

Darwin often complained during those years of both the magnitude of his project and the frailty of his health: "My health has been lately very bad from overwork.... My work is everlasting", he wrote on 16 April 1858.[9] Then, suddenly, in mid-June, came the manuscript of a young naturalist, Alfred Russel Wallace, advancing a theory about how varieties tend to form species. Darwin immediately perceived its striking proximity to his own theory and feared that "all [his] originality, whatever it may amount to, will be smashed".[10] Darwin now found himself trapped in something of a moral dilemma, caught between his desire to avoid being forestalled and his loyalty to Wallace, but also between his dream of completing the larger version of his work and the necessity to hastily provide his readers with a shorter version of his theory.[11] He asked both Charles Lyell, in his role as Lord Chancellor, and Joseph D. Hooker to help him find a fitting solution. A reading of different manuscripts was arranged during a meeting of the Linnaean Society held on the 1st of July. This involved Wallace's paper together with an extract from Darwin's 1844 essay and a part of the sketch Darwin had sent to Asa Gray in September 1857. The following August, the texts were published in the *Proceedings* of the Linnaean Society. Soon after this Darwin realised that, now that his ideas had been made public, he had no choice but to write an extended version of his ideas, providing a detailed account of the evidence he had gathered in support of it. Soon after his baby Charles had succumbed to scarlet fever, Darwin fled with his family to the Isle of Wight. Here, he began to think that a series of papers in the journal of the Linnaean Society would do the job of setting out his theory more fully: "I pass my time by doing daily a couple of hours of my Abstract, and I find it amusing and improving work". To which he adds the ironic comment, "It seems a queer plan to give an abstract of an unpublished work; nevertheless I repeat I am extremely glad I have begun in earnest on it".[12] Then, by mid-October, Darwin "expect[s his] abstract will run into a small volume, which will have to be published separately".[13] In the end, *On the Origin of Species* was written in just over a year: by April 1859, Darwin had negotiated with the publisher John Murray and a copy of the whole manuscript, drafted by the schoolmaster of the village of Down, was sent to the printer; by mid-June, Darwin received the first galley-proofs, to which he brought significant corrections; the proofs were finally corrected on 1 October 1859. Darwin wrote in his diary: "Finished proofs (thirteen months and ten days) of Abstract on *Origin of Species*".[14]

Darwin always quite bluntly presented his book as an "abstract", inviting all the expected correlative implications this brought with it. Usually, its status as an "abstract" is perceived as being a positive quality, since it resulted in Darwin maintaining a clear line of argumentation. Darwin himself, however, certainly saw it as a failing. Having amassed hundreds of pages of material, Darwin (strictly to avoid being beaten to the post) decided at first to publish an abstract of his book under the quite cumbersome title of "*An abstract of an essay on the origin of species and varieties through natural selection*".[15] In the first edition of the book, *On the Origin of Species by Means of Natural Selection, or the Preservation of Favoured Races in the Struggle for Life*, Darwin makes constant reference to a "longer work" that he was planning to complete.[16] The original *Natural Selection* manuscript was pushed aside (it would eventually be published posthumously in 1975) to make room for other writings, not to mention to allow time for Darwin's involvement in the numerous debates sparked by the *Origin*. Instead, Darwin dedicated considerable time to carefully reworking the 1859 text, gradually turning his *Origin* into a book of many versions. During Darwin's own life, no fewer than six successive editions were published by John Murray.

Around 75 per cent of the book underwent modification, with its overall length increasing by about one third in total. All of these changes are documented in Morse Peckham's *Variorum* text—a book which transforms how the reader looks at the *Origin*, presuming that the reader actually manages to navigate its almost unreadable maze of additions and corrections. The first edition came out in November 1859 [a]; the second in January 1860 [b]; the third in April 1861 [c]; the fourth in December 1866 [d]; the fifth in August 1869 [e]; and the sixth in February 1872 [f]. The bracketed letters [a], [b], [c], [d], [e], [f] are used by Morse Peckham for quick reference to the various editions and I will follow his system here.[17]

Almost every aspect of the *Origin* changed during its long life, right down to its date of birth: "October 1st, 1859" in [a] became "November 24th, 1859" in [d]. Even the title changed, the initial *"On"* disappearing with the publication of [f]. Some critics go so far as to suggest that dramatic modifications even changed the overall meaning of the *Origin* as well as the role that Darwin attributed to natural selection in evolutionary processes. But, unfortunately, within the immensely wide surface of Darwinian literature, those modifications remain today as what Darwin might have called "a grand and almost untrodden field of inquiry".[18] Although we know what kind of misprints distinguish the first printing from its successors ("speceies" on page 20, line 11), little has been written about the conceptual or theoretical evolution of the *Origin*. Two notable exceptions here are the works of Peter Vorzimmer and Helen P. Liepman. Peter Vorzimmer's study, *Charles Darwin: the Years of Controversy. The* Origin *of species and Its Critics* (1972), takes as its focus the period between the first publication of the *Origin* (1859) and Darwin's death (1882). Vorzimmer identifies "some fluctuation" in Darwin's thought on the role of natural selection: while in 1859 Darwin may have thought that he had effectively demonstrated the validity of natural selection,

> however, by the end of his life in 1882, Darwin had become, under critical attack, increasingly frustrated by his inability to prove to the satisfaction of fellow scientists that the selection process was the sole or, in some cases, the principal agent of evolutionary development.[19]

Throughout his book, Vorzimmer aims at showing that Darwin was muddled by his own poor conception of inheritance. Helen P. Liepman also documents an evolution in Darwin's text: while the first alterations, in the main, support the theory of accumulation of modifications by natural selection, the last two editions grant more leeway to non-selective forces.[20]

In spite of these contributions to the literature, it is still difficult to get a grasp on the issues and concerns that are *really* (i.e. theoretically, intellectually, conceptually, literally, or even socially) at play in and behind the various editions. The present chapter deals with the following questions: do any major changes of doctrine occur in the *Origin*, especially regarding the role of natural selection in the transformation of species, in the years from 1859 to 1872? In other words, is the 1859 edition of the *Origin* better conceived of as a culmination or as just a starting point? Can the later modifications Darwin made to the original edition all more or less be reduced to the categories of corrected typos, stylistic improvements, and annotation? After all, as Robert C. Stauffer has noted, "the *Origin of species* was, I believe, unique among Darwin's published books and formal scientific papers in appearing without a single footnote".[21] But in spite of its lack of footnotes, the *Origin* also, in some respects, constitutes a work of compilation, as Darwin himself wrote to Hooker while gathering data for his "big book" on species.[22]

As a compiler, Darwin stacks facts in the appropriate place. On the other hand, Darwin's modifications to his text betray much deeper revisions. If so, in what way do these different versions of the text display different theoretical choices that can all nevertheless be attributed to Darwin? Once in print, the waverings of Darwin-the-man offer a plurality of Darwin-the-texts that can in turn be pitted against each other.

Six different editions

Following Darwin's death, the sixth edition was widely considered to be the best, the one that represented Darwin's final word on the origin of species. But in 1950, Cyril D. Darlington published what he described as "the first reprint of the first edition, the only changes being in punctuation". In his foreword, Darlington claims that, due to Wallace's essay (which gave Darwin "the shock of his life"), Darwin drew up the *Origin* as a short abstract:

> Freed from the mass of learned references, it gave the gist of the story; it gave what everybody wanted to know. The strange thing is that the *Origin of species* as we now see it in its first edition still gives the gist of the story. The foundations that it laid have provided the foundations of a large part of science just as Newton's foundation did nearly two hundred years earlier.[23]

Darlington was undoubtedly the first to bring the first edition back to the fore and he went to a great deal of effort in order to justify his claim that it was this edition that should be read ahead of any other.[24]

Although Darlington's advice was not heeded by Julian Huxley for the Mentor Edition (1958),[25] in 1964 Ernst Mayr, the Harvard ornithologist serving as director of the Museum of Comparative Zoology, published a *facsimile* of the first edition of the *Origin*, in which he stated: "When we go back to the *Origin*, we want the version that stirred up the Western world, the first edition".[26] Mayr's reprint, however, made no mention of Darlington's prior undertaking. Due to the wide distribution of the Harvard reprint, it is undoubtedly Mayr we have to thank for popularising this novel way of considering that, given the history of the *Origin*, the first edition is still undoubtedly the best one to read. After Mayr, it became accepted that the post-1859 editions clearly drifted away from Darwin's original intentions.

There have been very few exceptions among subsequent editions in English to Darlington and Mayr's considered opinion.[27] Gillian Beer chose to print the second edition, since it might itself "be considered to be a reprint", or rather "Darwin's immediate urgent response to the private and public reception of the

Origin".[28] But as for the third to sixth editions, it seems that their fate in the archives has been definitively marked as "Out of print". As Beer puts it, these later editions "move the text increasingly far away from many of Darwin's initial arguments in an attempt to assimilate the responses and critiques of his contemporaries, particularly palaeontologists and physicists".[29] Much more now than a mere historical detail, the variation of the *Origin* throughout its successive editions is acknowledged as a theoretical problem.

Why, as Darlington and Mayr claim, should the first edition be given preference? Darwin described his book as "one long argument".[30] It seems, however, that through the process of revision this "one" argument became less and less visible; that Darwin made his case for natural selection "less concise and pithy", as Joseph Carroll put it,[31] and more equivocal and diffuse. The first edition, asserts Penguin editor John Burrow, "presents in many ways a more clear-cut and forceful version of Darwin's theory" than the five later editions: "Darwin weakened his argument in an attempt to meet criticisms".[32] After 1859, Darwin allowed himself to be led along the wrong path by critics whose objections belied commitments to either barren hypotheses or archaic frames of mind. And even in those cases where the book was actually "bettered" by a modification, either clarified or sharpened, it seems that, still, we have no genuine need for those later versions that lack the freshness of the first edition. Only a few short weeks separate versions [a] and [b], something Gillian Beer advances as her explanation for choosing to reprint the latter. John Murray, Darwin's publisher, was more than 250 copies short for the orders received during his autumn sale and so he asked Darwin, then undergoing hydropathic treatment in Ilkley, Yorkshire, if he would revise his text.[33] The "Ilkley edition" contains only 7 per cent of the total variants found in editions [b] to [f].[34] Darwin described it as "merely a reprint of the first with a few verbal corrections and some omissions", or "only Reprint; yet I have made a few important corrections".[35] As for the third edition [c], major changes include an "Historical sketch" where Darwin acknowledges the achievements of his predecessors, along with a postscript on Asa Gray's favourable review, suggesting that Darwin wished to reconcile his theory with natural theology—but both ended up disappearing again by the fourth edition. Within the body of the third edition, almost half of its specific changes (which, in sum, amount to 14 per cent of the total changes across all editions) appear in Chapters IV and IX. These include Darwin's reaction to some major reviews (by Owen, Bronn, and Harvey). The fourth edition [d] represents 21 per cent of the total revisions. It gives titles to many previously unnamed sections and makes allowances for new discoveries and new objections (Falconer).

The fifth edition [e], comprises nearly 30 per cent of the total changes and a more than 20 per cent increase in length. It is noteworthy on at least two accounts: for the introduction of Herbert Spencer's famous expression "survival of the fittest" and for the responses it provides to certain important objections (like those of Fleeming Jenkin).

The sixth edition [f] is usually regarded as the last and was edited just after the publication of the *Descent of Man* (1871), which had sparked renewed and expanded interest in Darwin's works. Designed for a wider audience, it was smaller and cheaper than its predecessors (about half the previous price) and also included a glossary, a key element in modifying the cultural status of the book, in that it fully assumed its new popular readership target audience.[36] A new chapter on "Miscellaneous Objections to the theory of natural selection" (Chapter VII), consisted of parts taken from Chapter IV cobbled together with additional material whose chief aim was to rebut St. George Mivart's attacks (whose *Genesis of Species* was also published in 1871). The major issue with [f] resides in Darwin's suspected "Lamarckianism": had Darwin now diminished the role of natural selection and put more stress on "other means of modification", such as habit, use and disuse of parts, direct effect of external conditions, etc.? With the 1872 edition, Darwin's modification of the *Origin* came, at last, to an end. He corrected one more final printing that was issued in 1876, but this contained only the most minor of changes.

The influences that motivated Darwin to change his text are numerous. Ernst Haeckel is responsible for the introduction of the phrase "phylogeny, or the lines of descent of all organic beings".[37] Other publications, such as Herbert Spencer's *Principles of Biology* (1864), influenced Darwin's maturing views on the question of variations.[38] Beyond such savant input, he was also tremendously affected by reviews and critiques and these, too, became powerful incentives in the ceaseless process of revising the text. Furthermore, he used later editions as a means to correct certain blunders: his figures on the denudation of the Weald, for instance, disappear from [c].[39]

In 1983, Gillian Beer suggested a different interpretation of how Darwin modified the *Origin*: she claimed that "the multivocality of Darwin's language reached its furthest extent" in [a], due to him relying on expressivity rather than rigour.[40] The expressive qualities of [a] allowed space for an exuberance and multiplicity of significations that were not to be permitted in later editions: Darwin, then combatting misunderstandings, felt the need to be more consistent with his lexical palette. Yet, it seems that the more Darwin modified his book, the more intertextuality he added, the more the layers of multiple meanings became inextricably entangled.

The *Origin* evolved considerably between 1859 and 1872 and its various editions provide perfect ammunition for those wishing to uncover a variety of Darwins. In the following sections, I introduce several domains where fluctuations in Darwin's thinking can be spotted: natural selection, variation, progressive evolution, Lamarckism, religion. To begin, I will lay out just an overview of how each theme changed through the six editions of the *Origin*. We will then return to these various domains in greater detail insofar as they constitute the main focus of later chapters.

The power of natural selection

The lynchpin of the dramatic alterations the *Origin* underwent from its first edition to its last must surely be the role attributed to natural selection and the description Darwin gave of his central concept.[41] He was especially cautious not to treat the causal mechanism he had discovered as some kind of magical wand, that is, as an all-powerful principle that would account for any and every structure. We have already seen how, in the case of the New Zealand debates, Darwin was mocked for referring (in [a]) to a black bear seen swimming "with widely opened mouth, thus catching, like a whale, insects in the water". In this first version, his stated position was that he could "see no difficulty in a race of bears being rendered, by natural selection, more and more aquatic in their structure and habits, with larger and larger mouths, till a creature was produced as monstrous as a whale".[42] In [b], he added "*almost* like a whale" to the first sentence and simply deleted the second one, likely in the wake of ridicule from Charles Lyell. Despite this, Darwin was not to be left in peace as regards the bear/whale story. Before Samuel Butler and the Bishop of Wellington's arguments involving discrepancies between the first and third editions, Richard Owen, in his 1860 review, had already compared the first two versions and then violently attacked Darwin's attempts to amend his own text. To Owen, the wording of [b] was no improvement on [a]: where [a] was vague, [b] was clear testimony to Darwin's cowardliness.[43] Owen was indignant: in modifying his text, Darwin changes only details instead of revising and clarifying the global argument. Owen was criticising Darwin for his miserly addition of a single word "*almost*", because, in reality, the problem posed by the bear/whale controversy was a much more serious one: what *can* actually be effected by natural selection?

From early on in the *Origin*'s existence, Darwin had to explain that he did not conceive of natural selection "as an active power or Deity" (*c*85, *Var* 165). The great failing of the term "Natural selection" is that it tends to personify Nature: but one should never forget that it is a metaphorical expression. Darwin made efforts to be more accurate, recognising that "several writers [had] misapprehended or objected to the term natural selection".[44] Darwin bluntly confessed that a large portion of the criticisms levelled at his beloved term were well-founded: "In the literal sense of the word, no doubt, natural selection is a misnomer". Later, in the fifth edition, Darwin would go so far as to call it "a false term".[45] Nature does not *literally* select, only metaphorically. Darwin even tried to explain the merit in the paradoxical character of the term "natural selection" to Heinrich G. Bronn, the German translator of the *Origin*: "its meaning is *not* obvious and each man could not put on it his own interpretation"; furthermore, the expression connects both "variation under domestication and nature".[46] Personifications also cloud around the term "Nature" itself. In one addition, Darwin suggests how it is "difficult to avoid personifying the word *Nature*" and he firmly reminds us that by "Nature" he means only "the aggregate action and product of many general laws, and by laws the sequence of events as ascertained by us".[47]

In [d], Darwin deals with the issue of natural selection gradually producing "utter and absolute sterility" between two species.[48] This point is crucial for the question of "the origin of species", or what Darwin terms in 1869 "species in process of formation" (*e*318). Darwin's sentence was quoted by his son Francis Darwin in a polemic against George Romanes.[49] In 1886, Romanes claimed that natural selection was not sufficient to explain the formation of interspecific steri-lity and that another process was required, a process he called "physiological selection". This suggestion entailed a huge debate on the "Darwinian" character of Romanes' hypothesis.[50] Did Darwin's sentence support Romanes' physiologi-cal selection? Or, on the contrary, did it destroy Romanes' claim to innovation? Through this example, we see that [a] is not the only version to have played an instrumental part in the reception of Darwin's ideas. Darwinian scholarship, including that of Darwin's own son, has always been eager to dig into the various layers of the *Origin*, to find the unexpected gold nugget that supports their own—or ruins others'—claims either to being Darwin's true heirs or, perhaps a greater prize, to being themselves *original*.

Moreover, it seems that over time Darwin became more committed to isola-tion as a necessary condition for speciation, a point he had previously denied.[51] Here, the German naturalist Moritz Wagner undoubtedly played a role.[52] By 1868, Wagner was convinced that isolation was the all-important factor in accounting for the origin of species, even terming it "the law of migration" (*Migrationsgesetz*). Darwin partly acknowledges the importance of this factor, while clearly stating that he "can by no means agree with this naturalist, that migration and isolation are necessary for the formation of new species". On the contrary, if "an isolated area be very small ... the total number of the inhabitants will be small; and this will retard the production of new species through natural selection, by decreasing the chances of the appearance of favourable individual differences" (*e*120, *Var* 196). In this instance, we see Darwin trying to appease a critic all the while standing firm on his key argument (the importance of natural selection).

Survival of the fittest

Bringing these successive restrictions and re-elaborations to bear on the concept of natural selection, it is the primacy of useful variation or of advantage as the fundamental factor in evolution that emerges with most clarity. As Alphonse de Candolle remarked, "people believed in this evolution without understanding how it could have been effected. Selection came along to provide an explanation for it, *or at least for the fixation of changes once they had been effected*".[53] Wallace, however, pointed out that Darwin tended to hide these differences in level, since the *Origin* superposes two meanings for "natural selection": on the one hand, natural selection designates a process at work, "survival of the fittest", and, on the other hand, it describes the result of that action, that is, the changes effected through the survival of the fittest.

Wallace had crossed out "natural selection" in his own copy of the *Origin* and substituted it with "survival of the fittest".[54] In a letter dated 2 July 1866, Wallace vivaciously argues that "natural selection", although crystal clear to some readers, is nonetheless a stumbling block for many others.[55] He went on to suggest the following: since the English term *selection* seems to introduce an "intelligence" into the process, and since, furthermore, "natural selection" is equated with "preservation of favoured races" (which can plausibly lead to progress), there was no objection to identifying Darwin's "natural selection" with Herbert Spencer's *"survival of the fittest"*. "Survival of the fittest", rather than preservation, was the perfect way of avoiding "natural selection" coming to signify a personification of nature through analogy with the intelligent intervention of the breeder. Affirming the metaphorical character of the expression *natural selection*, Wallace indicated that nature "does not so much *select* special variations, as *exterminate* the most unfavourable ones".[56] Ultimately, it seems that Darwin bowed to Wallace's arguments, as evidenced by the introduction of *survival of the fittest* into the fifth and sixth editions of the *Origin* (1869 and 1872 respectively).[57]

As Samuel Butler pointed out, Darwin appears to grant the reader free licence to take "survival of the fittest" as an equivalent to "natural selection", though perhaps this should in turn be taken to mean "the fertility of the fittest".[58] Butler further expounded that, while this "survival" or "fertility" may have been the *"sine qua non* for modification", it in no way implied that it could in itself be interpreted as "especial *means of modification*". Besides this, Darwin remained ambiguous on the status of natural selection and its potential causal relation to modifications: the proclaimed equivalence between *natural selection* and *descent with modification*, terms Darwin alternated in designating his system, only helped prolong confusion on this point. Taking it as a "means" of modification, natural selection is an agent, a genuine cause of variation; but, on the other hand, it is not the cause of individual modifications and explains neither generic nor specific differences: "the individual differences given by nature, which man for some object selects, must of necessity first occur".[59] As long as the cause of individual variations was not forthcoming, Darwin was "giving us an '*Origin of species*' with 'the origin' cut out".[60]

It is fair to argue that it did make a discernible difference whether readers of the *Origin* read the book with or without Spencer's phrase included. A good example of this is Samuel F. Clarke's work on larval cannibalism in salamanders. Published in 1878 under the title "An interesting case of natural selection", it concludes with the words: "It has been a very interesting case of natural selection, by survival of the fittest. All the weaker individuals being destroyed and actually aiding the stronger ones by serving them as food". The work was immediately reviewed in *Nature* as a case "illustrating survival of the fittest", triggering further comments on the same subject by A. Crane in a subsequent issue.[61] "Natural selection" soon became a pretext for scientific gibbering on "survival of the fittest".

Leaps, sports and slight variations

Another issue to be considered is whether Darwin's views on variation also evolved from 1859 to 1872. In the first edition, minute variations are necessary to the Darwinian process, but can one argue that progressively, albeit reluctantly, Darwin did eventually accept "sports" into his account? Did Darwin move from a theory where variations are minute and continuous to a more saltationist vision?

Thomas Henry Huxley, both in private letters and in published reviews, was very critical of the principle "*Natura non facit saltum*".[62] In light of this, Darwin made two changes in the second edition: the phrase "that old canon in natural history" becomes "that old but somewhat exaggerated canon"; and he simply deletes the sentence referring directly to *Natura non facit saltum*.[63] On similar points, "infinitesimally small inherited modifications" becomes "of small inherited" in the third edition.[64] Regarding continuous variation, and in spite of all the hesitations often attributed to him, Darwin quite categorically dispensed with the objection that new species can appear by saltations and, in [f], reaffirmed his commitment to continuous variation, saying that a "conclusion, which implies great breaks or discontinuity in the series, appears to me improbable in the highest degree".[65]

[c] Devotes specific attention to "various good objections" raised by H.G. Bronn.[66] Bronn thought the Darwinian theory required "that all the species of a region" should be "changing at the same time". Darwin always considered this to be an unnecessary supposition: "it is sufficient for us", he replies, "if some few forms at any one time are variable". Bronn also remarked "that distinct species do not differ from each other in single characters alone, but in many"; and he asked how it comes to be that "natural selection should always have simultaneously affected many parts of the organisation?" To this, Darwin replied that "probably the whole amount of difference has not been simultaneously effected; and the unknown laws of correlation will certainly account for, but not strictly explain, much simultaneous modification".[67] So, correlation of growth and a new emphasis on the laws of variation were, as we can see here, Darwin's go-to answer for many objections.

William Harvey also raised influential objections on the problem of saltations and monstrosities. For Harvey, the origin of species cannot be causally explained until the origin of variation is better understood. As he writes to Darwin,

> until however something more is known of the inciting causes of the *Variation and Correlation of Organs*, which in nature ever go hand in hand, I can only regard Natural Selection as one Agent out of several;—a handmaid or wetnurse—so to say—but neither the housekeeper, nor the mistress of the house.[68]

The nature of variation impacts on the status of natural selection. [e], for instance, contains an important (and much discussed) attempt to respond to the British engineer Fleeming Jenkin's "able and valuable article in the 'North British Review' (1867)".[69] Jenkin objected that there are absolute limits to variation, that a new form of a living entity would be swamped, and that the Earth is much younger than assumed by Darwin. Jenkin's review increased Darwin's attentiveness to the problem of variations: "I did not appreciate how rarely single variations, whether slight or strongly-marked could be perpetuated".[70] Jenkin had taken the case of "a highly-favoured white" who, shipwrecked on an island, fails to "blanch a nation of negroes". Darwin interpreted this as a convincing case *against* single variations and rethought the respective roles of individual differences (occurring in several organisms) and of single variations (rare and discontinuous forms of change). Jenkin's review appears to have led Darwin to relax his emphasis on natural selection, for instance when he writes: "The conditions might indeed act in so energetic and definite a manner as to lead to the same modification in all the individuals of the species without the aid of selection".[71] Darwin de-emphasised sports so as to place more emphasis on the normal range of variability. Indeed, Darwin's insistence on the individual level is easily perceived in many of the additions and modifications to [e]. At the beginning of Chapter IV, "an endless number of strange peculiarities" becomes "peculiar variations", before finally ending up as "slight variations and individual differences".[72]

Where the first edition reads: "A large amount of inheritable and diversified variability is favourable, but I believe mere individual differences suffice for the work", the fifth states: "A great amount of variability, under which term individual differences are always included, will evidently be favourable".[73] Other examples of Darwin's focus on individual variation can be found in the fifth edition.[74] The limits and scope of variation are obviously of great concern to him, and he does tackle the question of whether "many changes would have to be effected simultaneously": Darwin confesses that "this could not be done through natural selection", but, confidently relying on his 1868 work on *Variation*, he considers it to be a superfluous condition.[75]

Progress

Did Darwin accept the idea of tendencies in evolution, specifically tendencies towards a degree of superior "highness" in organisation? As Thomas Kuhn put it, readers of the *Origin* were not so much bothered by the general idea of evolution ("by no means the greatest of the difficulties the Darwinians faced"): "neither the notion of species change nor the possible descent of man from apes" were really a problem, since "evidence pointing to evolution, including the evolution of man, had been accumulating for decades".[76] What stood in the way of general approval of Darwin's ideas was "the abolition of [the] teleological kind of evolution", the fact that "the *Origin of Species* recognized no goal set either by God or nature".[77] In the light of Darwin's theory, terms like "evolution", "development", or "progress", seemed to have become self-contradicting.

Progress was clearly a matter of concern for Darwin. As early as [b], in the summary of Chapter IV, he adds a complement to the sentence "This principle of preservation, I have called, for the sake of brevity, Natural Selection", with the remark that natural selection "leads to the improvement of each creature in relation to its organic and inorganic conditions of life"—a sentence further developed in [c]: "and consequently, in most cases, to what must be regarded as an advance in organisation".[78] In the discussion on geological succession,

Darwin also adds that the best definition of highness is greater division of physi-ological labour and, consequently, that natural selection "will constantly tend" to make later forms "higher" than their progenitors.[79]

In [c], a whole new section is added: "On the degree to which organisation tends to advance".[80] Here, Darwin clearly distinguishes "highness" from "pro-gress". On the one hand, he asserts that:

> If we look at the differentiation and specialisation of the several organs of each being when adult (and this will include the advancement of the brain for intellectual purposes) as the best standard of highness of organisation, natural selection clearly leads towards highness.[81]

On the other hand, Darwin plainly denies progressive development, stating that "natural selection includes no necessary and universal law of advancement or development—it only takes advantage of such variations as arise and are bene-ficial to each creature under its complex relations of life".[82]

A notable feature of the sixth edition is the belated entrance of the actual word *evolution* into the text of the *Origin*. Previously, the *Origin* contained only the word "evolved", and even this appeared only in the closing words of the book. In [f], "evolution" occurs eight times and Darwin makes explicit reference to "the theory of evolution through natural selection". Generally speaking, the prior absence of this term has been attributed to two distinct reasons: first, to avoid confusion with Herbert Spencer's use of the word; and second, to avoid confusion with its embryological meaning of "development".[83] As a result, it may seem like the eventual introduction of the term is a sign that Darwin was prepared to accept more confusion on these issues. But it might just as well be contended that, by the 1870s, the term *evolution* was simply much more com-monly in use and that Darwin was, in fact, making things clearer rather than murkier for his readers.[84]

(Lamarckian) inheritance?

The formidable—though hazy—question of "Darwin's Lamarckianism" must certainly be the primary argument for avoiding later editions of the *Origin*, as was made clear in C.D. Darlington's reprint of the first edition. But the question is an anachronistic one, stemming mainly from Weismann's refutation of "Lamarckian inheritance", and one that I think should be avoided by any means possible. "Lamarckian" mechanisms generally include what Darwin called "use and disuse" or "direct effect of external conditions". Are these factors or forces thrust into a more active or efficient role in the last edition compared with the first?

In the *Origin*, Darwin clearly states that natural selection can aid in dispens-ing with Lamarckian explanations.[85] Nonetheless, the obvious signs of Darwin's affinities for "Lamarckian" factors are numerous, even in the first edition. The direct effects of environment on organisms are admitted in several passages, such as: "we must not forget that climate, food, &c., probably produce some slight and direct effect".[86] Then, on "habitual action" that becomes inheritable: "I think it can be shown that this does sometimes happen".[87] The various instances of Lamarckian themes in the *Origin* seem to coalesce in a key sentence with which Darwin closes the introduction: "Furthermore, I am convinced that Natural Selection has been the main but not exclusive means of modification".[88]

We will return to the issue of what those other "means of modification" are in later chapters. For now, it suffices to say that Darwin is chiefly referring to the action of a changing environment. Some have claimed to document a shift of emphasis in the role devoted to natural selection in the fifth and sixth editions. They advance that, up to the fifth edition, Darwin had modified his text such that it increasingly supported the theory of accumulation of modifications by natural selection, whereas in the last editions it was non-selective forces that he began to bring into play. Loren Eiseley was particularly vocal in stating that

> a close examination of the last edition of the *Origin* reveals that in attempting on scattered pages to meet the objections being launched against his theory the much-laboured-upon volume had become contradictory [...] [T]he last repairs to the *Origin* reveal ... how very shaky Darwin's theoretical structure had become. His gracious ability to compromise had produced some striking inconsistencies.[89]

However, all things considered, the evidence presented by supporters of the "Lamarckianisation of Darwin" hypothesis is rather frail. Can we claim that Darwin gave "extra stress to the direct action of the conditions of life" just because, where [a] reads, "We should remember that climate, food, &c., probably have some little direct influence on the organisation", Darwin went on to change *little* into, "some, *perhaps a considerable,* direct influence"[90]? It is impossible to conclude from this alone that Darwin had significantly changed his views. There are some obvious and probably not insignificant changes, such as, in Chapter I: "Habit also has a decided influence", which ends up becoming "Changed habits produce an inherited effect";[91] or the next sentence: "In animals it has a more marked effect", which would finally read, "With animals the increased use or disuse of parts has had a more marked influence". But evidence of the contrary can also be mustered, like this sentence from the revised Chapter VII where Darwin refers to "the inherited effects of the increased use of parts, and perhaps of their disuse" being "strengthened by natural selection": "How much to attribute in each particular case to the effects of use, and how much to natural selection, it seems impossible to decide".[92] The emphasis on a "tendency to vary in the same manner" is strong, in both this latter passage and others such as:

> There can also be little doubt that the tendency to vary in the same manner has often been so strong that all the individuals of the same species have been similarly modified *without the aid of any form of selection.*[93]

This tendency to vary certainly leads Darwin away from selection. But on the other hand, in opposition to Mivart's belief that species change requires "an internal force or tendency", Darwin is very clear:

> *there is no need*, as it seems to me, to invoke any internal force beyond the tendency to ordinary variability, which through the aid of selection by man has given rise to many well-adapted domestic races, and which through the aid of natural selection would equally well give rise by graduated steps to natural races or species.[94]

In fact, it is the Darwin manifest in [f], more than any earlier incarnation, who may stand accused of making an all-powerful operator of natural selection. This was already the case in [c],[95] but it was the constancy of the accusation which provoked clear changes in various passages. For instance, [a] states that "species have changed and are still slowly changing by the preservation and accumulation of successive slight favourable variations".[96] From [b] to [d], this same passage reads: "that species have been modified, during a long course of descent, by the preservation or natural selection of many successive slight favourable variations". In [e], the end becomes: "a long course of descent, chiefly through the natural selection of numerous successive, slight, favourable variations". But [f] adds to this again, stating that selection has been

> aided in an important manner by the inherited effects of the use and disuse of parts; and in an unimportant manner ... by the direct action of external conditions, and by variations which seem to us in our ignorance to arise spontaneously.

Does such an addition reflect some important change in Darwin's own perception of his theory? Strikingly enough, Darwin took specific pains to refer these later additions back to the closing sentence of the 1859 version introduction; that "natural selection has been, the main but not the exclusive means of modification".[97] Why then see ruptures when Darwin himself indicated continuities? From Darwin's own perspective, nothing had changed: he was simply trying to clarify a point that he had been making since the beginning but one which had been constantly overlooked. It is only from an "ultra-Darwinian" vantage point (one that equates Darwin with natural selection and *only* with natural selection) that Darwin can be accused of having changed his theory. But even as "Darwin-the-text" was being modified, "Darwin-the-man" seemed to be quite at ease with these changes of inflection in the perception of his theory and the interpretation of his book.

Reference to the Creator

Another example is Darwin's commitment to religion. As Darwin's name has become a flagship for atheism, one can legitimately wonder whether the *Origin* became more and more critical of theological language over the course of its modifications. Some, eager to reconcile science and religion, hope instead that Darwin's text became more and more accommodating to a Creator. What then is the truth of the matter?

[a] Refers to "the laws impressed on matter by the Creator" and twice includes the Pentateuchal verb "breathe", suggestive of a supreme Power.[98] These two instances of "breathe" were never removed, and, in fact, [b] goes a step further by adding "by the Creator" after them. In fact, though, Darwin was altogether very uncomfortable with the verb "breathe". The addition of "the Creator" in the final paragraph remains unchanged throughout the subsequent editions. But Darwin seems to have vacillated with the other occurrences: [c] deletes "Creator" again and replaces it with a long development on the principle of natural selection.[99]

Other additions are ambiguous. Darwin had stated that "two individuals must always unite for each birth"[100]—to which he added a parenthesis in [b]: "with the exception of the curious and not well-understood cases of parthenogenesis". What motivated this biological addition? Gillian Beer cleverly suggests that this may represent a theological addition in disguise, with parthenogenesis standing for Virgin birth.[101]

The truth is that Darwin was editing for a theological audience, as is suggested by the transformation of the phrase "natural selection will account for the infinite diversity in structure and function of the mouths of insects" into "natural selection acting on some originally created form will account for".[102] He also grants pride of place to any hint of support coming from theologians, like Charles Kingsley,[103] and, on the reverse side of the half-title page, he added a third quotation, placed between those from Whewell and Bacon: it comes from Joseph Butler's *Analogy of Revealed Religion*, a trusted resource for theologians of the time. With its reference to an "intelligent agent", it might easily be read as part of a strategy of Darwin's to mollify religious readers.[104]

What does the prism of the six editions reveal?

So, what have we learned by looking at the *Origin* through the prism of its six editions? First, it supports our methodological claim that it is a worthy undertaking to read the *Origin* as a problematical object in itself, without looking to later developments of Darwin's work or thinking. For instance, and notably enough, the "provisional hypothesis of pangenesis", developed in the 1868 *Variation*, never found its way into the *Origin*, apart from one brief mention in [e, 196] that no sooner disappeared again in [f]. On a general level, no full-scale revision of the structure of the argument was ever attempted. The creation of an additional chapter in [f] was aimed only at gathering scattered objections into one single body, while also considerably lightening Chapter IV.

As to the evolution of Darwin's views, while depicting Darwin as an exclusive selectionist is inaccurate, it is hard to claim that Darwin became *more and more* Lamarckian. Darwin just increasingly stresses the power of variations acting in unison with the power of natural selection. The laws of variation constrain natural selection. They also entail a refutation of any pan-utilitarianism; within the organism, not every structure is necessarily perfectly adapted or useful. Darwin's lifelong interest for the laws of variation led him to deep consideration of some so-called "Lamarckian" factors, such as the effects of changed conditions of life. Darwin envisioned channels leading from the environment into the organism, such that adaptive change could happen independently of natural selection. As a result, the *Origin* brings evidence that Darwin was an evolutionary pluralist.

But, as is often the case in Darwinian processes, minute and sometimes imperceptible modifications may have dramatically altered the meaning of the whole.[105] Darwin's interest in variation led him to put progressively more stress on other factors which "aid" natural selection: the fulcrum of the *Origin*'s argument shifts gradually from the confines of Chapter IV into the realm of Chapter V. During this process, publications such as 1868's *The Variation of Animals and Plants* show his attempts to gather a considerable amount of reliable raw material in which to place his confidence. It seems that the main incentive for changes was Darwin's wish to address objections and criticisms that had been brought. However, this shift of focus from natural selection to other means, including the laws of variation, does not equate to a "Lamarckianisation" of his thinking.

As for Darwin, the study of the causes of variation was part of his original project, his "big book" on the variation of species and natural selection. It may be claimed that some early chapters of *Natural Selection*, now missing from the manuscript, were included *ne varietur* into *Variation*.[106] In this view, *Variation* is to be taken as part of an initial three-fold plan: after the publication of the "abstract" (i.e. the *Origin* understood as a preliminary announcement), Darwin was resolved to start anew, first with the book on variation, and only then planning to proceed to natural selection and the "problem of the conversion of varieties into species", the "main subject of my second work" as he claimed in *Variation*, before finally progressing to "a third work" devoted to seeing to what extent natural selection "will give a fair explanation" of "several large and independent classes of facts".[107] Given the general outrage sparked by the role of variation within the theory of the origin of species, one might also think that Darwin secretly hoped his study on variation would resolve some of the difficulties arising from the *Origin* by providing his readers with many facts only alluded to in the first chapters of the former. However, it turned out that this shift of emphasis, from the origin of species to the origin of variation, only amplified rather than decreased the difficulties. As a matter of fact, Darwin was never to publish *Natural Selection* … nor his "second work", nor indeed the third one.

Which is the best edition of the *Origin*?

We shall return now to the question underlying our use of the *Origin*'s six editions as an investigative prism: when various versions do enter into conflict, is there one version that should prevail over the others and, if so, on what account? In November 1901, when copyright of the first edition of the *Origin* had expired in Britain, John Murray printed the following note:

> Darwin's *Origin of Species* has now passed out of Copyright. It should, however, be clearly understood that the edition which thus loses its legal protection is the imperfect edition which the author subsequently revised, and which was accordingly superseded. The complete and authorized edition of the work will not lose copyright for some years.[108]

The insistence here on the last edition being the best is clear as day—a position that contradicts reading practices of the post-Mayr era, which favour the first printing over any later ones.

Murray's words should not, of course, be simply taken at face value. Huge financial interests were involved in his manoeuvre. However, their tone of warning may yet serve as a historical opening of the debate: on what grounds should one consider one version of the *Origin* to be *better* than the others?

An edition can be *better* than others according to various criteria:

- *Clarity:* if it sets forth the argument in a clearer, less convoluted way;
- *Thoroughness:* if it corrects typos and other obvious mistakes;
- *Contemporaneity:* if it provides the reader with new, up-to-date information.

The argument of clarity is key if one considers the *Origin* to be "one long argument".[109] On the other hand, the argument of contemporaneity may be unavoidable, given the historical contingency of scientific information: for instance, after fossils of the bird-reptile Archaeopteryx began to be uncovered in the early 1860s, the creature soon found its way into the *Origin*; the same happened with the concept of mimicry, following the publication of Henry Walter Bates' 1862 paper on the butterflies of the Amazonian region.[110]

Darwin clearly suggests that the argument of contemporaneity might well lead to the argument of thoroughness, since the text of the *Origin* depends on a state of scientific knowledge as well as on sources of information that may or may not turn out to be fully reliable. The arguments of clarity and contemporaneity may collide on some points, since clarity suggests simplification, while contemporaneity implies additional matter whose bulk may overshadow the general outline of the argument.

The argument of contemporaneity is tangible in those pages where Darwin leans more and more towards dogmatism. I will give but one example: the question of the dawn of civilisation.[111] The topic is cursorily evoked in the first chapter of the first edition: "Mr. Horner's researches have rendered it in some degree probable that man sufficiently civilized to have manufactured pottery existed in the valley of the Nile thirteen or fourteen thousand years ago". The reference disappears from the third edition, where it is replaced by a somewhat stronger statement:

Since the recent discoveries of flint tools or celts in the superficial deposits of France and England, few geologists will doubt that man, in a sufficiently civilized state to have manufactured weapons, existed at a period extremely remote as measured by years.

The fourth edition reads: "all geologists believe that man in a barbarous condition existed at an enormously remote period". As for the fifth edition: "all geologists believe that barbarian man existed at an enormously remote period". We see here how reliance on the very latest sources does not necessarily lead to more accurate statements. Not only does Darwin go from a tentative expression to a confident one, he also introduces a new figure, of a profoundly Victorian bent: "the barbarian man". In this case, historical generalisation and incorporation of new sources of information did not lead to any "improvement" of the text.

So, is there a *best* edition that readers should all acknowledge and whose text should prevail over all others? Or are readers left with the confusing maze the several versions entail, thus entitled to freely play one *Origin* against the other? The idea that Darwin was obsessed with providing a *better* version of his argument tends to downplay three important factors:

a   the role of external interventions (such as reviews) in pointing out faults in the *Origin*'s text;
b   Darwin himself was constantly editing his text, searching for perfection;
c   Darwin not only responded to reviews and edited typos, he was also constantly revising his own views and positions.

On the first two factors, readers and critics often constituted strong motivating incentives behind the rewriting process of various passages. Darwin often changed a word or two, suppressed a sentence here, added a development there, sometimes in the wake of some sneering comments or blatant misunderstanding. The third factor is much more central to the issue at stake here, that of several "Darwins" being played against each other.

Ernst Mayr, who played an active role in debates on the "best" edition of the *Origin*, stresses two

> extreme interpretations concerning the development of Darwin's theory of evolution ... both of them clearly erroneous [...] According to one, Darwin developed his theory in its entirety as soon as his conversion to evolutionism happened. The other extreme is to say that Darwin constantly changed his mind and that later in life he completely abandoned his earlier views.[112]

For Mayr, after Darwin started his notebooks on transmutation (1837), he quickly "adopted and rejected a series of theories", but then "he more or less retained the overall theory he had developed by the 1840s through the rest of his life ... without completely reversing himself". Hence, for Mayr, Darwin's modifications to the *Origin* matter little (they concern only secondary points), even though he also says that Darwin's

> statements on evolution in the sixth edition of the *Origin* (1872) and in the *Descent of Man* (1871) are remarkably similar to the statements in the essay of 1844 and in the first edition (1859) of the *Origin*, all contrary claims notwithstanding.[113]

So it is perhaps paradoxical that Mayr himself republished the first edition of the *Origin*, urging readers to come back to the original version, free of the "impurities" of later developments. Ultimately, in light of these contrasting views, readers of the *Origin* have to make difficult decisions regarding the dialectic of change and continuity in the Darwinian theory for themselves.

Contra Mayr 1982, I suggest it is probably the case that at least parts of the modifications do actually reveal Darwin's progressive revisions of his own initial position. But, Mayr's case persisting, if this is true, should we care? Probably the answer is "yes", if one is a biographer of Darwin or a historian of the development of evolutionary thought. But if one is an evolutionary biologist enquiring into the mechanisms at work in nature, an epistemologist interested only in the emergence of a "Darwinian paradigm", or a philosopher looking for the meaning of a consistent "Darwinism", then it is probable that Darwin's own hesitations are indeed of little import.

As we can see, the question of the *best* edition of the *Origin* is subject to diverse evaluations, as it refers us not only to Darwin's own understanding of his book but also to the expectations of his readers. Some readers—let's call them the Historicists—are keen to underscore possible changes in Darwin's own mind; while others—let's call them the Systematists—are interested only in the systematic aspect of the Darwinian philosophy of nature. These two categories of reader pay varying attention to the discrepancies between "Darwin" the historical man, subject to doubts and imperfections, and what is termed "Darwinism", i.e. a prescriptive set of expectations on what Darwin's thought is, should be, or should have been.

The *best* edition enquiry is therefore a highly debatable issue. Today, it seems to be the Systematists—be they biologists or philosophers—who are massively dominant. The work of historians of science is ambiguous on this point. They have eagerly published Darwin's papers, from the early notebooks to the long manuscript on species that Darwin was working on when he was interrupted by Wallace's paper and precipitated into the hasty composition of what would become the *Origin*. However, many of these historians seem to embrace the idea that what matters most is, in fact, the emergence of Darwin as a consistent selectionist figure, very close to the person dearest to the heart of evolutionary biologists. Few historians have paid attention to the historical changes of the *Origin* and Peckham's *Variorum* edition might well be dubbed "the greatest Darwinian study never read", in part because of the book's somewhat unreadable format, but partly also because of a persistent lack of general interest in the history of the book itself. Even fewer scholars have paid attention to the various disputes that arose from misunderstandings between readers referring to these diverging and at times conflicting versions of Darwin's *opus magnum*. However, the fact that the *Origin* did constantly evolve played a major role in the fact that readers of the book ended up pledging their voices to differing views of what Darwin, or rather "their" Darwin, *really* thought.

Notes

1 The ideas presented here were first published as "The Evolution of the Origin (1859–1872)". In Michael Ruse (ed.), *The Cambridge Encyclopaedia of Darwin and Evolutionary Thought*, Cambridge University Press, 2013, pp. 158–164. The present version has been modified and expanded.
2 "Barell organs" (17 January 1863), in Butler 1923.
3 Ellegård 1958, pp. 238–241.
4 It is the subtitle of DNS.
5 See Darwin's Pocket Diary for 9 September 1854.
6 Darwin to W.D. Fox, 19 March 1855, CCD 5 288.
7 Darwin to W.D. Fox, 3 October 1856, CCD 6 238; Darwin to C. Lyell, 10 November 1856, CCD 6 256. Interestingly enough, Darwin applied the same expression when speaking of his two volumes on *Variation*. See Darwin to T.H. Huxley, 7 January 1867.
8 See Darwin to Asa Gray, 5 September 1857, CCD 6 447.
9 Darwin to W.D. Fox, 16 April 1858, CCD 7 70.
10 Darwin to Charles Lyell, 18 June 1858, CCD 7 107.
11 Loewenberg (1959, p. 7) has beautifully expressed how Darwin's reaction to Wallace's contribution involved more than anguish over priority. See also Loewenberg 1957.
12 Darwin to J.D. Hooker, 30 July 1858, CCD 7 141.
13 Darwin to J.D. Hooker, 12 October 1858, CCD 7 168.
14 Diary entry for 1 October 1859.
15 See enclosure in Darwin to Charles Lyell, 28 March 1859, CCD 7 270.
16 See *Origin* 1859: on the "abstract", p. 1 and 481; on Darwin's "future work", p. 192 and 216.
17 Now all versions are available on websites such as darwin-online, edited by John Van Whye.
18 *Origin* 1859, p. 486.
19 Vorzimmer 1972, p. xviii.
20 Liepman 1981.
21 DNS, pp. 1–2.
22 Darwin to Hooker, 2 May 1857, CCD 6 389.
23 Darlington 1950, xvii.
24 *Ibid.*, xix–xx.
25 See Huxley (J) 1964. This edition is a reprint of the sixth edition. No indication is given by the publisher.
26 Mayr 1964, p. xxiii.
27 One of these exceptions is the Dent and Sons edition (London, 1928), reprinted in 1971 with an introduction by L. Harrison Matthews, whose text reproduces the sixth edition.
28 Beer 1996, p. xxix.
29 *Ibid.*, p. xxix.
30 *Origin* 1859, p. 459.

31 Carroll 2003, p. 76.
32 Burrow 1985, p. 49.
33 Dixon and Radick 2009.
34 *Var* 773.
35 Darwin to Murray, 2 December 1859, CCD 7 411, to Gray, 21 December 1859, CCD 7 440.
36 On Darwin's personal involvement in settling the lower price of the sixth edition, see Desmond and Moore 1991, p. 582.
37 *Origin* 1869, p. 515, *Var* 676.
38 Compare *Origin* 1859, pp. 131–132 with 1869, 165–166, *Var* 276.
39 *Origin* 1859, p. 287, *Var* 484.
40 Beer 1983, p. 38.
41 We have seen in the introduction how some of the finest Darwin scholars, Moore 1979 or Gayon 1998, have striven to identify a core theory in Darwin, one that does not vary along the multiple editions.
42 *Origin* 1859, p. 184, *Var* 333.
43 Owen 1860a, 519.
44 *Origin* 1861, p. 84, *Var* 164.
45 *Origin* 1861, p. 85, 1869, p. 93, *Var* 165.
46 Darwin to Bronn, 14 February 1860, CCD 8 83.
47 *Origin* 1861, p. 85, *Var* 165.
48 *Origin* 1866, p. 311, *Var* 444.
49 See *Nature*, vol. 34, 2 September 1886.
50 On this, see below Chapter 6.
51 See, for instance, *Origin* 1859, p. 105.
52 On Wagner, see *infra*, Chapter 6.
53 Candolle 1873, p. 15 (emphasis added).
54 Beddall 1988.
55 Wallace to Darwin, 2 July 1866, CCD 14 227.
56 *Ibid.*
57 Cf. *Var* 164 (#13:e) and *Var* 145 (#15.I:e) where Spencer's expression is said to be "more accurate" and "sometimes equally convenient".
58 Butler 1879, p. 351.
59 Darwin 1861, p. 84; *Var* 165 (#14.3:c).
60 Butler 1879, p. 363.
61 See Clarke 1878, the review in "Biological notes", *Nature* 19 (no. 477), pp. 155–156 (19 December 1878) and Crane 1879.
62 See 23 November 1859, CCD 7 39.
63 Check these changes in *Origin* 1859, p. 194 (*Var* 361) and p. 210 (*Var* 383).
64 Compare *Origin* 1859, p. 95 with 1861, p. 100, *Var* 185.
65 *Origin* 1872, p. 201, *Var* 264.
66 *Origin* 1861, p. 139, *Var* 230.
67 *Origin* 1861, p. 140, *Var* 231.
68 Harvey to Darwin, 24 August 1860, CCD 8 322.
69 *Origin* 1869, p. 104, *Var* 178.
70 *Ibid.* On the importance of Jenkin, see Vorzimmer 1972, pp. 121–126; Gayon 1998, pp. 85–102.
71 *Origin* 1869, p. 105, *Var* 179.
72 *Origin*, resp. 1859, p. 80; 1869, p. 91; 1872, p. 62, *Var* 163.
73 *Origin* 1859, p. 102; 1869, p. 117, *Var* 192.
74 *Origin* 1869, p. 94 and 104, *Var* 166 and *Var* 178.
75 *Origin* 1869, p. 225, *Var* 342.
76 Kuhn 1970, p. 171.
77 *Ibid.*, p. 172.

78 *Origin* 1860, p. 127, 1861, p. 144 and *Var* 271.
79 *Origin* 1860, p. 336, *Var* 547.
80 *Origin* 1861, pp. 133–137, *Var* 220–226.
81 *Origin* 1861, p. 134, *Var* 222.
82 *Origin* 1861, p. 135, *Var* 223.
83 Bowler 1975.
84 E.g. Ruse 2013, p. 12: "Evolution was a pseudoscience. Darwin changed that. Evolution (at the least) was now an accepted fact; it is common sense."
85 *Origin* 1859, p. 242.
86 *Ibid.*, p. 85.
87 *Ibid.*, p. 209.
88 *Ibid.*, p. 6.
89 Eiseley 1959a, p. 242. This passage is cited as an epigraph to Vorzimmer 1972 (p. 1).
90 *Origin* 1859, p. 196, *Var* 363.
91 *Origin* 1859, p. 11, 1872, p. 8, *Var* 83.
92 *Origin* 1872, p. 188, *Var* 253.
93 *Origin* 1872, p. 72, *Var* 179.
94 *Origin* 1872, p. 201, *Var* 264.
95 *Origin* 1861, pp. 84–85.
96 *Origin* 1859, p. 480, *Var* 747.
97 *Var* 747.
98 See *Origin* 1859, p. 484, 490, 488 (*Var* 758).
99 *Origin* 1861, p. 519, *Var* 753.
100 *Origin* 1859, p. 96, *Var* 185.
101 *Origin* 1860, p. 96. Beer 1996, p. xxiv.
102 *Origin* 1859, p. 436; 1860, p. 435, *Var* 678.
103 *Origin* 1860, p. 481, *Var* 748.
104 Darwin's own method had recourse to analogy—namely that of artificial and natural selection (Jon Hodge, personal communication).
105 See Vorzimmer 1972, p. 270.
106 See DNS, p. 11.
107 See Darwin 1868, vol 1, p. 5 and p. 9. This view is endorsed by Stauffer in DNS, pp. 11–13.
108 Quoted in Kohler and Kohler 2009, p. 335.
109 *Origin* 1859, p. 459.
110 Bates 1862. *Origin* 1866, pp. 367 and 503. On Bates and the importance of mimicry, see Gayon 1998, pp. 178–196; Kimler and Ruse 2013.
111 Compare *Origin* 1859, p. 18, with 1861, p. 18, 1866, p. 19, 1869, pp. 19–20.
112 Mayr 1982, p. 410.
113 *Ibid.*

2 Diffracting Darwin's title—the prism of translations

The first thing any reader encounters, even before opening the book, is its title. The title of Darwin's book, in particular, is as ample as it is mysterious:

> *On the Origin of Species by Means of Natural Selection, or the Preservation of Favoured Races in the Struggle for Life.*

Over the course of the six editions this title remained practically unchanged. Only with the sixth edition was the initial "On" dropped.

Darwin's title is a mysterious object. It is worth our while to read it with intertextuality in mind. For Gillian Beer, "the title is in polemical contrast with Chambers's insistence on *Vestiges of the natural history of Creation*. Vestiges are remnants, surviving fragments of a primordial creative act. Darwin's enterprise is history, not cosmogony".[1] Personally, I remember being struck by the imitation game the *Origin of Species* set off: a series of *spin-off* titles popped up, from St George Mivart's *Genesis of Species* to Theodor Eimer's *Origin of Species by Means of Inheritance of Acquired Characters*, or Edward Drinker Cope's *Origin of Genera* or "*Origin of the fittest*"—the title he gave to his collection of essays. But when I began to think in terms of imitation, *Origin of Species* also struck me as a possible variation on a certain conventional title type. Indeed, Darwin's title may itself be an oblique reference to other works, like Thomas Vernon Wollaston's *On the Variation of Species*, published in 1856. The *Origin* often refers to the studies of Wollaston and Wollaston's *Variation of Species* happens to be dedicated to none other than Charles Darwin himself, an author who had contributed to the progress of "zoological geography" and "to whom we are indebted for so much valuable information concerning the natural history of various portions of the world".[2]

Another book on species that was published in 1859 is Dominique Alexandre Godron's two-volume synthesis simply entitled *De l'espèce et des races dans les êtres organisés* (*On the Species and Breeds of Organised Beings*).[3] Of course, it is out of the question here to claim that Godron, then Dean of the Science Faculty in Nancy, could have in any way influenced Darwin in his choice of title: Godron's book had no public impact and was by no means destined to enjoy the glorious fate of the *Origin*.[4] But its concomitant publication indicates two things: first, that book titles referring to species were quite common among naturalists of the time; and second, that the two books, Darwin's and Godron's, in spite of their differences, would be often compared by other contemporary naturalists.[5]

In recent biological literature, Darwin's title has attracted no fewer commentaries and critics. Ernst Mayr was particularly vocal in expressing the general suspicion that the book known as the *Origin of Species* does not in fact constitute a theory of "the origin of species". Mayr made his bewilderment on the matter known in his study of the history of biology: not a single one of Darwin's fourteen chapters gives "the origin of species"! As Mayr comments:

> The origin of species—that is, the multiplication of species—is such a key problem in Darwin's theory of evolution that one would surely expect it to be the exclusive subject of one of the fourteen chapters of the *Origin*. But this is not the case. The discussion of speciation forms part of Chapter IV, a chapter dealing primarily with the causation of evolutionary change and with divergence. When reading this chapter, one is struck by the insufficiency of the analysis.[6]

In the 1990s, it became something of a cliché to state that the *Origin* did not answer the question it itself raised. In 1991, Helena Cronin voiced this dramatic conclusion: "But, in the midst of such success, there was one problem that remained just outside [Darwin's] grasp. It was—poignantly—the problem of the origin of species".[7] Stephen Jay Gould, with a just touch of irony, joined the discussion in 1992: "Darwin did wrestle brilliantly and triumphantly with the problem of adaptation, but he had limited success with the issue of diversity—even though he titled his book with reference to his relative failure: the origin of species".[8] In 1993, the philosopher Elliott Sober went as far as claiming that Darwin's book should have been titled: "*On the unreality of species as shown by natural selection*".[9] The biologist Steve Jones, reader in genetics at University College London, bluntly claimed that if Darwin were to publish his best-known book under its existing title today, "he would have been in trouble under the Trades Descriptions Act", because "if there is one thing which the *Origin* is not about, it is the origin of species. Darwin knew nothing about genetics".[10] Daniel Dennett shared the same stance in 1995: "Notice that Darwin's summary does not mention speciation at all. It is entirely about the adaptation of organisms, the *excellence* of their design, not the diversity".[11]

As a result, the title has been a difficult pill to swallow, which is precisely why scholars like Jeff Wallace have called for a process of "defamiliarisation" since new incentives now exist "to prise Darwin's title out of its self-evidence and look at it afresh" in order to expose its ambiguities.[12] The present chapter takes up this issue of "defamiliarisation" but examines the question through a prism comprised of the *Origin*'s various translations into other languages.

A book in translation

Soon after its publication in English,[13] Darwin's book was translated into several European languages; the second British edition was translated into German in 1860, the third into French in 1862 and Italian in 1864. The first Dutch and Russian editions were based on the second edition, the Swedish from the fifth. The fourth French edition (by Col. Moulinié) was based on the fifth and sixth English editions (for the first and latter halves, respectively). By their reflections on the original English wording, this accidental community of translators help us penetrate the extent to which Darwin's beautiful title is nevertheless most obscure. Through them, we come face to face with the question: what did Darwin actually mean when referring to "the origin of species"? This aspect of our quest into the meanings of Darwin's theory intersects with the growing field of scholarship known as "translation studies in science", what David N. Livingstone has called the "geographies of reading", emerging from the recent understanding that "thoughts routinely travel the world in textual form".[14] As Nicolaas Rupke has put it, mapping the geographies of textual reception discloses "the constitutive significance of place in the production of the various meanings that became attached to even a single work".[15] This new emphasis on the spatiality of knowledge focuses heavily on various cultural preoccupations. For instance, French and German reviews of Alexander von Humboldt's *Essai politique sur le Royaume de la Nouvelle-Espagne* focused on his contribution to an accurate cartography of Mexico, while British ones dwelt more on the aspect of commercial strategies.[16] Similarly, the conclusions of a lifelong opponent to Darwin's theory, such as John McCrady (1831–1881), may have had their source in, for example, support of slavery in South Carolina and the Confederate states, while the lectures William Travers (1819–1903) gave on Darwin in New Zealand are coloured by the idea that Darwinian evolution explains the triumph of white colonial settlements, etc.[17] In this light, each reading of the *Origin* emerges as "shaped by the cultural politics of [its] interpretive community", as Livingstone phrased it. Geographies of knowledge show how "the coming together of texts and readers ... is a moment of creativity in which meaning is made and remade",[18] but the interpretations they result in tend to pass through the lens of social or political agendas. The wonderful project of "locating" theories, or the "interweaving of place, politics and rhetoric", often delivers the message that the reception of scientific ideas is subject to distinct ideological or cultural determinisms.

In contrast to the above methods for introducing place and time into the history of science, the present chapter focuses on languages: I want to show how the challenge of translating English words into other languages helped shape the reception of Darwin's ideas. In travelling, the *Origin* did not only encounter new cultures: it had to be reframed within new conceptual spaces, which in turn led to a range of conceptual interpretations.[19] Usually, this is where scholars roll out variations on the Italian paronomasia *Traddutore, traditore* (translator, traitor). This has been especially the case in the French example, with the works of

Clémence Royer who was blamed for many mis-readings and biased interpretations.[20] My own claim is that, far from "polluting" the original meaning of the text, these translations actually generate more clarity as to what Darwin's title actually means. I dubbed this "the Dogmatix miracle", in reference to Obélix's little white dog in the English translation of the comic book *Astérix le Gaulois*.[21] Having to render the name of the dog "Idéfix", an original pun on the French expression *idée fixe* (fixed idea or obsession), the translators came up with "Dogmatix". The English name is probably even better than the French original, since it not only preserves the meaning of the original pun but even adds to it, by combining the reference to the doggedness (and "dogginess"!) of the character with the allusion to the notion of dogmatism. This example perfectly illustrates how translations are often able to preserve all the assets of the original word or phrase and then even surpass it. It is my contention that the translations of the *Origin* performed just such "Dogmatix miracles": preserving the meaning of Darwin's original wording while also adding new clarity to it by disentangling the range of meanings contained in the original English terms.

Far from being the occasion to mete out more lukewarm platitudes on "*traduttore, traditore*", closely examining the translations proposed for the title of Darwin's *Origin* creates a prism that reveals just how difficult it is to grasp the book's true nature as regards, notably: the pertinence of the question it poses (*the origin of species*), the signification of what is taken to be its central concept (*natural selection*), and the manner in which the latter is connected to the *struggle for existence*. The simple act of examining how the title has been translated illuminates a whole host of what would otherwise be hidden interpretation problems. It can even alter our impression of the work as a whole.

And yet, the *Origin* shines with a sort of self-evidence that results in its being embraced without question. Its success, the speed with which the first edition sold out, the way contemporary discussion immediately re-centred around the origin of species; all of this adds up to make 24 November 1859, the date of the work's first publication, a historical event of the highest order.

Darwin's title through the prism of its translations

Now, let us imagine that we have no idea what the book is about, that we are completely ignorant to the book's actual subject. Let us apply the translation prism and see how an apparent unity (Darwin's title) breaks into a multiplicity of colours and meanings.

Passing the title through the translation prism reveals new dimensions that were enclosed within it; a new terrain full of discrepancy and controversy unfolds. One person will say that the title alone is a distillation of the entire theory, another will maintain that the title *Origin of Species* is nothing but a "misnomer".[22] After all, such an uncomprehending stance towards the title cannot be said to be entirely affected; if anything, it is to be expected when we remind ourselves that it is not transparent even for English speaking readers and that several competing versions of the title exist in other languages, each representing its own divergent interpretation.

As an initial example, let us first compare two German versions of it:

Bronn 1860	*Ueber die Entstehung der Arten im Thier- und Pflanzenreich durch natürliche Züchtung, oder Erhaltung der vervollkommneten Rassen im Kampfe um's Dasein*
Seidlitz 1871	*Ueber den Ursprung der Arten in Folge von Naturauslese, oder die Erhaltung der begünstigteren Rassen im Lebenskampf*

Immediately, we see that the German offers two distinct terms for practically every word of the English title, the most significant differences being the following:

Origin	*Species*	*Natural selection*	*Favoured*	*Struggle for life*
Entstehung	*Arten*	*natürliche Züchtung*	*vervollkommneten*	*Kampfe um's Dasein*
Ursprung	*Arten*	*Naturauslese*	*begünstigteren*	*Lebenskampf*

Difficulties in interpreting the title of Darwin's book are certainly not unique to the German language, however. Quite noticeable differences are also to be remarked in French translations of the title, from Clémence Royer's loose 1862 adaptation of it to the more literal versions proposed by later translators (Moulinié or Barbier). What choices must French translators make?

Royer 1862	*De l'origine des espèces, ou Des lois de progrès chez les êtres organisés*
Royer 1866	*De l'origine des espèces par sélection naturelle, ou Des lois de transformation des êtres organisés*
Moulinié 1873	*L'Origine des espèces au moyen de la sélection naturelle ou la lutte pour l'existence dans la nature*

We can see that, while Moulinié's translation may be, word for word, closer to the original, it still grants itself some liberty: for example, it omits the reference to "favoured races", thereby highlighting the "struggle for existence", which it treats as being practically equivalent to the origin of species. More radical in their divergence, the translations proposed by Royer forego the concept of natural selection and introduce the notion of "*lois de progrès*" (laws of progress) in 1862, and then "*lois de transformation*" in 1866.

This brief initial schema on its own is sufficient to instil perplexity, and yet it covers only a fraction of the difficulties. Every component part, up to and including the small word "means" and the simple "or" connecting the two parts of the title, has proven itself to be cause for commentary. Hence, the title, in its entirety, can be passed through the translation prism such that each new version reveals new potentialities that were all already implicit in Darwin's original words.

On the *Origin* ...

Bert J. Loewenberg noted in 1959 that "the word 'origin' was never used in its sense of 'beginning'; it always implied changes in the development of life-forms already in existence".[23] However, as Jeff Wallace added: "neither Darwin's text nor his culture could allow him to limit the word thus".[24] This came to be especially true when non-Anglophone readers endeavoured to understand the book and render it in their native tongues.

If we take this point seriously, then any reading of the *Origin of Species* should begin with due consideration of the hesitation shown by Darwin's contemporaries as they endeavoured to penetrate the meaning of the word "origin". Although the French and Italians contented themselves with the term "*origine*", it may be that this apparently commodious solution actually obscured some genuine problems. These can be drawn to the surface in two ways. First, by noting that the term "*origine*" has been used as a French and Italian translation for other key English terms, such as "descent". And, second, by looking to what the title became in languages where a cognate term like "*origine*" was not an option. When it comes to translating "origin", the German language proves itself a convenient prism for diffracting the spectrum of colours hidden within the white light of the title *On the Origin of Species*. Indeed, in German a real work of translation was required (in that an equivalent had to be found within a distinct lexical system) in the absence of a simple transliteration option: the choices made throw light on what the question of origin fundamentally determines.

Georg Bronn proposed *Entstehung* for the first German translation, a work he undertook using the second English edition of the *Origin* (1860).[25] The term *Entstehung* denotes the process of some thing appearing, and this can in turn be retro-translated by terms like *emergence* or *appearance*, as opposed to *vergehen* (pass away, vanish). In translating "origin" by *Entstehung*, there is a shift from the "origin" of species to their "mode of emergence". Is this the same thing?

Whatever advantage this term may have is best measured against its potential competitors, notably the term *Ursprung* that is also to be found in the literature on Darwin.[26] *Ursprung* denotes the source, the starting point, the originating point (contributed by the prefix *Ur-*). Victor Carus, the second German translator of Darwin's work, tasked with correcting Bronn's version, proposed a modification of the title which would replace *Entstehung* with *Ursprung*. Other sources inform us that Darwin was not satisfied with Bronn's translation, so one would tend to trust Carus and acquiesce to his interpretation of the title. Indeed, Darwin himself confessed that *Ursprung* might well be a better term than *Entstehung*: however, on the very same point, he remained strictly perfunctory and, in fact, categorically forbade the new term to be used, so as not, he claimed, to bewilder the public by presenting them with the same book under different titles.[27] But, behind the pretext of wishing to avoid confusing the "public" by having two titles for the one book, was Darwin perhaps not just improvising an indirect, though polite, argument for refusing Carus' proposition, one which he may have deemed to be simply incorrect? The intransigence with which he rejected *Ursprung* seems to indicate a difference of opinion that goes beyond simple circumstantial concerns. Following this view, *Ursprung* would not be the correct translation for origin and *Entstehung* must be seen as the preferred choice, whatever qualities or faults Bronn's translation may have besides, and this before any consideration of the theoretical twists to which it subjects Darwin's text.

The search for the primordial "source" does not cover the full depth of the word "origin". The verb "to originate" points primarily to a mode of production, an action with a certain power, an efficient cause. This point is decisive if we are to discern whether the book does or does not provide an answer to the question its title poses: if "origin" is *Ursprung*, primary source, then the question being asked bears first on primordial ancestors (original prototypes, the very first progenitors of life on earth), and then, from that point, on the genealogical order of derivation linking the extinct forms unearthed by palaeontology to the present forms discovered by natural history. If, on the other hand, "origin" is *Entstehung*, then the question no longer bears on the origin of life or the genealogical tree but on the very mechanism or process at work within nature itself. Within the foregoing view, Darwin's book would be seen to show how natural selection and the struggle for existence "produce" species, how these two concepts "originate" species, in the sense of *efficiently* producing them. Retro-translating from this interpretation back to English, *Origin of Species* might more properly be rendered as the *Production* or *Formation* of species.

This difference in meaning, seemingly trivial, nevertheless becomes quite apparent as one follows the trail of the various versions and retro-versions that were to appear. For instance, the French translation of a German work makes several references to a book by Darwin entitled *Formation des espèces*: this undoubtedly results from "origin" being translated into German as *Entstehung* and then this being translated into French as *Formation*, by either a very uninformed or perhaps very conceptually perceptive translator.[28] Another French book praises the title of the Darwinian opus—a title which in itself is "*une révélation anticipée*"—which the author cites as, "*Production des espèces à la faveur de la sélection naturelle ou à la faveur de la conservation des races, accomplies dans la lutte pour l'existence*".[29]

This short overview of the various meanings of the term *origin*, as evidenced by the variety of translations both in German or French, confirms Gillian Beer's suggestion that the title has to be read in full, precisely because its ordinary shortened form, *The Origin of Species*, "changes 'origin' from a process into a place or substantive".[30] Translations teach us that the shortened title *Origin of Species* has a tendency to turn *Entstehung* into *Ursprung* for German ears, *Production* or *Formation* into *Origine* for French ones.

... of Species ...

In the foregoing perspective, the book's professed aim would be to see how species are produced, Darwin claiming to reveal the origin of *species*, translated to French as *espèces*, to Italian as *specie*, and to German, most often, as *Arten* (more so than *Species*). Darwin's initial idea was to speak of "the origin of species and varieties". But his readers immediately raised objections of all sorts as Darwin complained to Murray:

> A friend objected to my title that [the] word "Varieties" ought to stand before "Species".—Another friend objected (but illogically) that "genera" and "orders" ought to be inserted. [...] This has led me to think that word "Varieties" had better be altogether omitted. The case of Species is the real important point; and the title, as now, is rather too long.—So if you do not object, I will omit the word "Varieties".[31]

The "case of species" alone constituted the "really important point".[32] Nevertheless, the concept of species is doubly problematic, on the one hand because its meaning (its definition) is obscure, even without considering the difficulties of its translation, and also because, on the other hand, pairing it with "origin" seems to create an oxymoron. Even if we knew exactly what *origin* was and what *species* was, the meaning of the compound form "*origin of species*" would still not be in any way clear. The ontological status of species has been a debated matter up to now: Darwin's work is often presented as being contradictory with species stability, leading to a form of "received view" that Darwin supported "species nominalism", but others have endeavoured to show that Darwin supported the *reality* of species taxa.[33]

The first reviews of the *Origin* countered Darwin by asserting that species is a "*naturae opus*", an idea based on aphorism 157 of Linnaeus' *Fundamenta botanica* ("As many species are counted as there were forms created in the beginning").[34] Richard Owen, for one, stressed that

> classification is the task of science, but species the work of nature': we believe that this aphorism will endure; we are certain that it has not yet been refuted; and we repeat in the words of Linnæus, "*Classis et Ordo est sapientiæ, Species naturæ opus*".[35]

In accordance with this, readers of this opinion present the history of the theory of evolution as a battle against the dogma of Linnaeus and Cuvier, which is to say, against the doctrine of immutability. However, the Linnaean aphorism, invariably taken to be a token of religious orthodoxy in favour of the fixity of forms and the symbol of metaphysical essentialism, tells us neither what a species is nor how to recognise one, we are told only how many are to be counted: it limits itself to indicating that "forms" were "created". But which forms were they?

In reality, the Linnaean creed supposes two dimensions—a fixist essentialism and a derivative continuism—between which Darwin creates a tension. To the first Linnaean *dictum* (aphorism 157 of *Fundamenta Botanica*) we can oppose an equally Linnaean continuity axiom: the *Natura non facit saltum* that concludes aphorism 77 of *Philosophia botanica*, which Darwin quotes several times. Aphorism 77 is also quoted as a natural conclusion to the Darwinian schema in a review by Francis Bowen,[36] and in another by Richard Simpson, who does not attribute it to Linnaeus but instead cites it as a traditional aphorism, a fabled and deceptive adage that leads Darwin down the path of some "mythological conclusion": "and man's descent will be traced, proximately perhaps, from an Adam the offspring of a baboon and, ultimately, from a monad, through a slug".[37] This same aphorism also led Clémence Royer to place Linnaeus among the continuist, and thus transformist, philosophies.[38]

Whatever was created in the beginning, was it *species*? By claiming that species (as is the case with varieties) have a common origin in a common type, that they result from the diversification of a primary form, we ultimately only push the fixed term back by one notch and make a claim of essentialism for genera rather than for species. If we look to the *dictum* of aphorism 157 and follow the botanist Alexander Braun's interpretation of it, Linnaeus himself seems to have recognised that all species of a genus form "originally only one kind (*ursprünglich nur eine Art*)", derived from a "common mother-form (*gemeinsame Mutterart*)".[39] What then is a species?

Considered as a taxonomic level, the species level sits between the (superior) genus or family level and the (inferior) variety level. The existence of *varieties* (or *races* or *sub-species*), coupled with the *variability* of the individual, poses the problem of species identification or determination. It should be noted that this definition was subject to many polemics: everybody "put forward their own [definition]", remarked a contemporary.[40]

To the Linnaean vision of species, whose double importance in the defence of species stability and ambiguity with respect to the very nature of species can be measured together, we can also add Buffon's conception, one which had an equally considerable influence. Buffon's significance to the definition of species stems from his redefinition of it as a line of interfertile individuals who produce offspring that are also fertile.[41] Darwin discusses these conceptions in Chapter VIII of the *Origin*, dealing with hybridism. His theory of descent with modification opens the way for him to blur any difference between species and variety and also to obscure the infertility barrier. On the problem of providing a definition for species, Darwin chooses to keep his silence. Notably, he highlights the deep divergences among naturalists as to whether a group of individuals does or does not form a species.[42] He also declares that "we shall have to treat species in the same manner as those naturalists treat genera, who admit that genera are merely artificial combinations made for convenience".[43] The beginning of Chapter II also feeds into conventionalist interpretations of species, particularly when Darwin writes, "I look at the term species, as one arbitrarily given for the sake of convenience to a set of individuals closely resembling each other".[44]

In this, Darwin confirms the indefinable nature of species, and his declarations are often interpreted as pointing towards a lack of solidity in the general nature of species itself. Darwin may simply have been trying to get around the problem of having to define species: is it not one of those terms so self-evident that attempts at definition can only obscure? Whether or not this is the case, his readers would not grant him this charity. Is it really possible, they asked conspiratorially, to claim to give "the origin of species" without even specifying what it is you're giving the origin of?[45] In such a view, the whole difficulty with the book, and with the Darwinian theory of "the origin of species" in general, would reside in the fact that "Darwin didn't clearly formulate for himself the meaning he attached to the word *species*". "Nowhere in his works", writes A. de Quatrefages "could I find anything precise on this matter: no trifling criticism to be able to level at an author who claims to have discovered the secret of the origin of species". Or, to put it more succinctly, species emerges here as "that organic unit to which the very ones who deny its reality never cease to return".[46]

It seems that, from the moment species have an "origin", the term species itself no longer means anything, as was famously stressed by Louis Agassiz (to whom we shall return below). "Species", stripped of its right to the status of essence, seems to retreat into either a nominalism or a conventionalism: the lesson of "Darwinism", far from revealing the secret of the origin of species, now boils down to the dissolution of species into its multiple component individuals. In reality, this "nominalism" of Darwin's is to be judged in two ways. On one side, erasing all "lines of demarcation" between species and varieties fulfils a critical function in countering the vision that assigns each one with its own specific "origin": specific acts of creation for each species and natural laws explaining only varieties. On the other side, the problem of fixing the species level and variety level is a profoundly non-conventionalist one, and Darwin himself goes so far as to state that "[this] classification is evidently not arbitrary like the grouping of the stars in constellations".[47] Therefore, it is possible to remain faithful to Darwin yet still believe that species are neither illusions nor arbitrary creations of the naturalist. This point was not always understood by readers of the *Origin*. For instance, the French theologian Hyacinthe de Valroger accused Hooker of an "inconsequential adhesion" to Darwin's creed on account that one and the same scientist could not both refute the fixity of species (by subscribing to Darwin's views), and yet believe that species are not illusions.[48]

Independently of what Darwin may have actually thought of species, the nominalist interpretation of his work was for a long time commonplace. Ernst Haeckel, for instance, proclaimed that, after a century of empty disputes in systematics—of trying to tell whether a form was "a good or a bad species, a species or a variety, a sub-species or a group"—the theory of descent had definitively resolved the question by affirming that these concepts "have no absolute meaning, but are merely stages (*Gruppenstufen*) in the classification, or systematic categories, and of relative importance only".[49] Similarly, Wallace, who opened his book *Darwinism* (1889) with a chapter entitled "What are 'species' and what is meant by their 'origin'", indicates that:

Before Darwin's work appeared, the great majority of naturalists, and almost without exception the whole literary and scientific world, held firmly to the belief that *species* were realities, and had not been derived from other species by any process accessible to us.

Obviously, all of this changed with the publication of the *Origin*, from which time

[the] whole scientific and literary world, even the whole educated public, accepts, as a matter of common knowledge, the origin of species from other allied species by the ordinary process of natural birth. The idea of special creation or any altogether exceptional mode of production is absolutely extinct.[50]

If Darwin did in fact give a meaning to the origin of species, then we are to believe that it was by giving up on the realism of the very concept of species, on the idea of "specially created" species. Clémence Royer went on to base a political and social philosophy on this nominalism. Opposing the principle of direct creations (where the notion of species is "a fixed and definite entity"), her philosophy aligns itself with "the formation of living creatures by secondary causes", where ascending and progressive evolution leads to "continual mutability" and where individuals are "the only realities, the only substantial entities". From this point on, species is mere "logical category, without reality, an entirely contingent resemblance of attributes having no essential link to the subjects in which they are manifest, and variable in each individual of each successive generation".[51] To the realist and fixed concept of species (which she associated with scholastic theologies, as well as with socialism, Platonism, and Christianity), Royer opposed nominalist individualism, which sweeps aside all exterior norms and which, alone, waves the flag of liberty. According to realist logic, species is a kind of must-be, a "divine idea that all individuals must realise" (i.e. an oppressive idea), while, conversely, "the unlimited employ" of freedom by individuals opens a much larger domain of possibilities, which nothing can limit a priori. In this, realist logicians are led to affirm that individual variation admits of no natural limit and, on the same score, they must also reject all transcendent and determinant norms. Royer's reading suggests the idea of an evolutionary success that creates itself out of the departure from predefined norms and is a factor of progress. This is precisely what "Mr. Charles Darwin has come to demonstrate today, in his beautiful book *On the Origin of Species*".[52]

Paradoxically, nominalism leaves the title devoid of any pertinence: why concern oneself with the origin of species if the concept of species in itself isn't actually any *thing* at all? In other words, if species is a non-evolutionary concept, and if, conversely, the mutation of forms is incompatible with the existence of species, then the Darwinian position, at least as it is presented in the title, runs the very real risk of being nonsensical. It would then have to be admitted that the term "species" was to be understood only in the common sense of "what the

naturalists or ordinary language habitually distinguish as different species",
regardless of what the real ontological status of the concept of species, or its
actual definition, or even definability, may be. This is the final lesson that
Darwin seems to draw: that by adopting his ideas, the naturalists "shall at least
be freed from the vain search for the undiscovered and undiscoverable essence
of the term species".[53] Despite this, the term "species" can still be used as a practical entity allowing for the description of natural phenomena.

... by means of ...

What is the nature of what Darwin "originated" under his own name? Darwin
never claimed ownership for "originating" the theory stating that species were
not created independently: "The only novelty in my work is the attempt to
explain *how* species become modified, and to a certain extent how the theory of
descent explains certain large classes of facts".[54] If this is the case, then Darwin's originality among all other theories opposed to the dogma of independent
creations would not reside in the idea of the descent of forms relative to each
other, but rather in the *how* of the operation he alone advanced. If the *Origin* is a
Production (an *Entstehung* rather than an *Ursprung*) and if *Species* is an entity
without a determined essence, formed by the coming together of several individuals in accordance with one point of view, then how would this "production"
take place? *By means of natural selection*, the title tells us. This inconspicuous
word "means" seems fairly harmless, but it has provoked its fair share of questions. Is it a simple connecting word, just like the "*through* natural selection"
Darwin used in his first intended title? Or is it a specific point whose interpretation is of primary importance, as Samuel Butler in particular suggested in his
Evolution Old and New (1879)?

> When Mr. Darwin says that natural selection is the most important "means"
> of modification, I am not sure that I understand what he wishes to imply by
> the word "means". I do not see how the fact that those animals which are best
> fitted for the conditions of their existence commonly survive in the struggle
> for life, can be called in any special sense a "means" of modification. "Means"
> is a dangerous word; it slips too easily into "cause". We have seen Mr. Darwin
> himself say that Buffon did not enter on "the causes or means" of modifica-
> tion, as though these two words were synonymous, or nearly so. Nevertheless,
> the use of the word "means" here enables Mr. Darwin to speak of Natural
> Selection as if it were an active cause (which he constantly does), and yet to
> avoid expressly maintaining that it is a cause of modification.[55]

This easily overlooked "means" may in fact constitute a decisive key to understanding Darwin's work, its originality, and its ambiguity: is natural selection
the cause that originates species or is it just a "means"? The existence of "other
means of modification" (*other* than natural selection) is a highly debated matter
to which we shall return in a further chapter.

For Gillian Beer, the emphasis in the title is on "means", as this term emphasises the dimension of time, movement, and forces involved in the process of origination that Darwin describes.[56]

... natural selection ...

The principal "means" advanced by Darwin, fully encapsulating the originality of his doctrine, is the concept of *natural selection*. Natural selection, we have seen, is the core of an orthodox reading of Darwin; that is, of Darwin-the-selectionist. Through the translation prism, we understand that "natural selection" alone is host to a variety of Darwins. In fact, far from being obvious, natural selection supports a great variety of interpretations.

German translators stumbled over the term "natural selection" and ended up with multiple equivalents. In Italian, the English *selection* was long translated as *elezione* and Giovanni Canestrini, the Italian translator of the *Origin*, held on to *elezione* right up until his last writings in the 1890s.[57] In French, must *natural selection* necessarily become "*sélection naturelle*"? Today, there seems to be no doubt over this. But does this lack of doubt carry any weight? In a review of Darwin's work from the September 1860 edition of *Magasin pittoresque*, the journalist annotated the expression with "*choix ou triage*" (choice or sorting). As for Clémence Royer, she preferred the expression "*élection naturelle*". This preference made her a suspect of all manner of intentions. It could even be considered that her use of the term "election" contained an element of heresy: in Darwinian territory, the "elected" are not the object of supernatural divine grace but are simply the best equipped in the "*concurrence vitale* (vital competition)"—the phrase Royer used to render "struggle for life".

In terms of the book's reception, everything played out over the fact that the term "*élection*" carries a more spiritualist signification than the term "*sélection*"; in choosing this term, Royer brought an "intelligence" into Darwin's thinking, introducing an imposter into the very framework of the *Origin*.[58] Owing to the same cause, Darwin was exposed to certain unfair criticisms, like those of Pierre Flourens, the permanent secretary of the Académie des Sciences, who accused Darwin of "personifying nature". Even more so than "*sélection*", "*élection*" relates to conscious and deliberate choice, even a certain measure of arbitration. Darwin obtains

> a metaphorical language which dazzles him and he imagines that the *natural election* he ascribes to nature would have huge, *incommensurable* (his word) effects that the feeble power of man does not have ... nothing stops him; he plays with nature as he pleases and has it do everything he wishes.

Flourens, through the old trick of truncated quotations, suggests that Darwin himself recognised the oxymoronic nature of the expression "*élection naturelle*", since "election" denotes an intelligent force while "natural" denotes precisely the absence of such intelligence. Selection, despite Darwin's protests, confers a sort of unconscious election onto nature, which is what Flourens criticised him for: "Either natural election is nothing, or else it is nature; but nature endowed with election, nature personified: the last error of the last century: the nineteenth century no longer fabricates personifications".[59] Here, Flourens may have been led astray by Royer's bad choice of term.

In reality, the English term "selection" carries just as much risk of personification as the French "*élection*", something seen by the fact that Darwin, as early as 1861, had to explain that he never meant natural selection "as an active power or Deity". On the contrary, "election" almost seems less open to such criticisms, as it is first and foremost a classical term from chemistry used for speaking about the *elective* affinities of inorganic matter.[60] It could also be argued that English is more prone to personifications than French. As Gillian Beer noted, "since English is an ungendered language one need only add a 'his' or 'hers' to turn a word into a personification".[61] Flourens, in this case, was only pointing at a difference between the French and English language and blaming Darwin (not Royer) for the faults of his own concept. Eventually, Royer did nevertheless switch "*élection*" for "*sélection*" in the second edition of her translation, although not without expressing her deep aversion to the term.

We have to ask whether Royer was to blame when choosing to use "*élection*" rather than "*sélection*". In other words, would it have been viable for her, in 1862, to translate "*natural selection*" with the French "*sélection naturelle*"?[62] The substantive "selection" could easily have been transliterated into French. But the issues are not the same with the verb "to select" and its derived adjective form "selected". In English, *election* and *selection* form etymological twins, or a doublet, directly modelled on the Latin: the verb *eligere (electus)* means to tear or dig out, and *seligere (selectus)* means to choose and put together, as is explained in any etymological dictionary (see Table 2.1). But, strikingly enough, although the term *eligo* had been directly translated into the very common French verb *élire*, there was no direct equivalent for *seligo*, and the derived adjective *selectus* was traditionally translated as *choisi*. There was no French verb available for expressing "to select". Accordingly, the easy solution was to render "to select" as "*choisir*" (to choose). This solution was retained by later translators of the *Origin*, such as Jean-Jacques Moulinié and Edmond Barbier.[63] However, this choice still entails the following problem: when a French reader encounters "*choisir*" in Darwin's text, there will be no immediate mental connection to the operation of

Table 2.1 The Latin doublet *eligo/seligo*

Latin	English	French
Eligo, electus	To elect, elected	*Élire, élu*
Seligo, selectus	To select, selected	• *Choisir, choisi?* • *Sélire?* • *Séliger?* • *Sélectionner?*

"selection". Furthermore, *choisir* is a very common verb and therefore loses the technical dimension of "selection". This is why Moulinié occasionally resorted to neologisms. For instance, in the "Introduction" to the *Origin*, he translated Darwin's expression "naturally selected" as "*naturellement conservé ou sélecté*", but in the next sentence, "selected variety" becomes "*variété ainsi épargnée* [saved/ rescued]".[64] As for Barbier, he simply avoided the verb and adjective forms as much as possible: his translations of the same two passages are "*être l'objet d'une sélection naturelle*" and "*variété objet de la sélection*".[65]

A young zoologist of invertebrates from Geneva, René Edouard Claparède (1832–1871), suggested that "*élection*" and the verb "*élire*" were an easy way to stay close to the English terms *selection, to select*, and *selected*, without the need for neologisms.[66] Claparède's initial suggestion was taken up by Clémence Royer in the first French edition of the *Origin*. Given the absence of any exact equivalent for *selection/to select*, the *élection/élire* solution does appear to be quite an elegant translation.[67] "*Élection*" might sound strange in the context of natural processes, but it is precisely *because* of this reason that it may have conveyed something of the technical dimension of the English term. Above all, it comes equipped with an equivalent family of words for the derived forms of *to select* (*selective, selected*). It seems likely that the foregoing lexical problem was reason enough in itself for Royer to choose "*élection*". Other attempts were made, but all proved unsatisfactory. Some suggested that the Latin verb *seligere* could be transcribed directly into French as "*séliger*".[68] Royer herself suggested that she may have been amenable to the neologism *sélire*. It would have sounded elegant, but it would also have been perplexing. The worst choice, or so Royer claims, would have been the ugly and ill-formed "*sélectionner*".[69] The term "choisir", retained by Barbier, precludes description of a neutral process of "selection" and, moreover, loses any proximity with *to select* and *selection*: precisely where Darwin had introduced the idea of an *unconscious selection*, it introduces the idea of some consciousness directing the process. For Royer, the affair was decided: one either had to decide to systematically translate *to select* with the neologism "*sélire*" (and not with *choisir*), or else return, as she did in the end, to speaking of *élection naturelle* in order to benefit from the flexibility offered by the French verb *élire*.

Before undertaking his German translation of the *Origin*, Bronn first wrote a review of it in 1860. There, he proposed "*die Wahl der Lebens-Weise*" as a translation for *natural selection*: turning it, literally, into "the choice of lifestyle (mode of life)".[70] The presence of the word "*Wahl*" translated the idea of a *choice* (rather than a *selection*), but even disregarding this aspect, it is clearly not the same as "natural selection". For Bronn, this expression simply conveyed a way of acknowledging that varieties are produced by differences in modes of nutrition, environments, and climates, as well as in many other factors. This is why Darwin complained heavily about this distortion of his views, and in compliance with Darwin's wishes, Bronn then proposed the translation *natürliche Züchtung*. The term *Züchtung* denotes both breeding and cultivation: to wit, it is adequate for describing the procedures applied to domestic species, though without making explicit the intervention of a *choice*.

But other translations were still possible. Georg Seidlitz instead proposed *Naturauslese*, which does evoke a "sorting" but leaves aside the explicit reference to breeding. As for Ludwig Büchner, he preferred to translate *selection* by *Auswahl*, a still more neutral term merely denoting a choice between several possibilities. What did he disapprove of in *Züchtung*, which seems to have the advantage of denoting "breeding (or cultivation)"? Büchner's French translator suggested that Bronn's *Züchtung* would imply the idea of an improvement or an amendment: in that case, Bronn would be translating *natural selection* as "natural amendment".[71] But can we really criticise him for this when Darwin's own suggestions went as far as *Adelung*, "ennobling"? In the new German translation, corrected by Victor Carus, Bronn's *Züchtung* became *Zuchtwahl*, a way of allying the two elements, domestication (*Zucht*) and choice (*Wahl*).

The inherent ambiguity of Darwin's own terminology turns on him as soon as it needs to be translated: every translation demands an interpretation, insofar as there exists no term in any other language which precisely matches the conceptual articulation of the original English. The solution, then, would seem to be to transcribe the "foreign" term into the target language: this is exactly what the French Darwinians ended up doing (by replacing *élection* with *sélection*), followed in turn by their German counterparts (with the expression *Selektionstheorie*). Since all translating implies interpretation, the meaning of "natural selection", which is in no way self-evident, was fated to be lost or displaced in translation. To avoid this, there was ultimately no choice but to settle for transliteration. From this perspective, it is absolutely legitimate to regret that Bronn's and others' choices convey a stowaway Lamarckism under the guise of "habits" where Darwin had placed only an analogy with the practices of breeders.

... or the preservation ...

But besides the question of whether translations can ever avoid importing parasitic significations, translations may well reveal the inherent difficulty of communicating the Darwinian conceptual lexicon into another language. Indeed, it would be foolish to think that "natural selection" is in itself clear to native Anglophone ears; Darwin's own editor, John Murray bemoaned the choice. Darwin, shocked, defended his expression, which was "constantly used in all works on breeding", although he did have to promise that he would complete the title with some subtext. He suggested, *"through natural selection or the preservation of favoured races"*.[72] Darwin acknowledged that the term was particularly difficult to understand and even stated that it was its very unintelligibility that made it appealing.[73]

Hence, while each translation proposal unlocks a different interpretative path of the Darwinian theory—none of them divorced from the very terms in which it was initially laid out—it would be wrong to think that the original version of the book is free of interpretative conflicts and that the English terms do not hide their own ambiguities which could hinder proper understanding of the Darwinian system. This is encapsulated in the fact that the book's title expounds on the expression "*by means of natural selection*" by adding the word "*preservation*". The title of the book specifies that selection signifies preservation, perpetuation, conservation (*Erhaltung*). The *Origin* explicitly states that the term *natural selection* was adopted only "for the sake of brevity" to denote "this principle of preservation", and in several letters from 1860 Darwin hints at regret over the choice of "natural selection", saying he should instead have spoken of "natural preservation".[74]

But shifting from *selection* to *preservation* of modifications has many implications. If the originality and meaning of the Darwinian theory ultimately resides in the concept of *natural selection*, if, moreover, its meaning is not immediately evident but demands a certain amount of elaboration, then this is enough to explain how many of Darwin's readers were left with the impression that he had merely given a new name to something long recognised. It explains and justifies, for example, Armand de Quatrefages saying of himself, "although without making use of the *word*, I have long professed the *thing*".[75] In this view, Darwin merely gave a new name to what was already known and, above, "the *word*" Quatrefages evokes comes from Augustin-Pyramus de Candolle's aphorism stating that "*nature is at war*". The paradox is that the term *selection*, if it be re-interpreted as "war", becomes almost flat, unfit for rendering what the interpretation says of it; conversely, *selection*, taken literally, runs the apparent risk of attributing a figurative "intelligent being role" to nature itself. Between flattening and overburdening, "*selection*" is forever problematic and Quatrefages proposed substituting it for the term "*elimination*", a word he deemed to be more precise. Another problem with Quatrefages equating Darwin's *selection* with De Candolle's *war* is that Darwin explicitly distinguished the two in his private notebooks. In an entry dated 28 September 1838, he writes: "Even the energetic language of ~~Malthus~~ DeCandoelle does not convey the warring of the species as inference from Malthus".[76] Quite paradoxically, Malthusian metaphors are more "war-like" than Candolle's own use of the word "war"!

Finally, perhaps Darwin's introduction of the *selection* concept is of no great importance, if indeed it is not the first time "at least part of the role it carries out in the general harmonies of the world" was understood. However, natural selection as elimination leaves us with no explanation for the genuinely creative processes in nature. As for the fact that this "elimination" paradoxically produces "harmonies", this, scoffed Quatrefages, is a lesson we had already been given in La Fontaine's fables.[77]

If natural selection is preservation, then it must be understood as preservation *of what?* Through this shift from *selection* to *preservation* of modifications, Darwin is seen to be changing direction: from the origin of species to the origin of variations. He finds himself necessarily guided towards a reflection on *what is preserved.* Hooker and Lyell, for instance, remarked that the term "natural selection" is not complete and they recommended the expression "variation and natural selection". For Lyell, to speak of natural selection as if it acts alone, neglecting to mention variations, was the same as assigning to it more work than it can do and the not carefully guarding against confounding it with the creative power to which "variation" and something far higher than mere variation *viz.* the capacity of ascending in the scale of being, must belong.[78]

The presence of initial variability imposes itself as the primary factor, the existence of variations upon which natural selection has the potential to intervene being only a secondary, ulterior factor.[79] The book's title stands accused of obscuring the fundamental role of variability, as though the insistence placed on natural selection prevented sufficient consideration of the material upon which this selection acted.

... of favoured races ...

Proceeding in this manner, if natural selection is interpreted as a dividing blade, then the creative dimension of the process must be explained and *natural selection* demands to be completed by *variation*. But another issue arises from the fact that Darwin does not say "natural selection" is "the preservation of useful variations". The title of his book instead speaks of "*the preservation of favoured races*".

The term "*races*" has attracted a lot of scholarly attention, since it seems to drag the whole debate on human evolution, ape ancestry, slavery, and racism into the Darwinian domain.[80] Reading the book through its title, as we are doing in the present chapter, we may want to further analyse Darwin's use of the term "race" and see whether or not it involves the human angle. Throughout the *Origin*, though notably in the first chapter, Darwin uses the term "races" many times, mostly in reference to garden plants or domestic animal breeds. The term "races" is often associated with the adjectives "domestic" or, although rarer, "geographical". My feeling is that his use of the term is as a synonym of "breeds", sometimes with taxonomical innuendoes, especially when the term is combined with others like "sub-species", "hereditary varieties" or "individuals".[81] Darwin maintains this use throughout his book. However, it is true that in three specific occurrences the term "races" is explicitly applied in relation to humans.[82] Hence, Darwin's use of the term "races" does not provide us with a clear-cut answer to the well-worn issue of whether or not the *Origin* was actually about "man".

How does the question break down when passed through the translation prism? In languages like French or German, "race" or "Rasse" is the most general term employed to describe domestic animal or plant "breeds". Hence, "races" is much more present in French or German translations than in the original English version. A quick word count reveals that the term "races" appears less than 50 times in the first edition of the English *Origin*, about 125 times in Bronn's German translation, and reaching an impressive 230 times in Royer's French version.[83] This difference of occurrences is chiefly explained by the fact that when Darwin writes "breeds" or even at times "kinds", the German version reads "Rassen" and the French version reads "races".[84] The Dutch version, by Tiberius Cornelius Winkler, also reaches a high count of "rassen" (about 275 occurrences) since he often uses the term as a translation of "varieties".[85] This very broad and loose use of the term "races" in languages other than English may have produced the effect of importing the concern for the evolution of human races into the *Origin.*

In stark contrast to this, one can conclude that Darwin's use of the term "race" was rather reasonable and restrictive: he used it mostly to designate domestic breeds, which explains the never-ending debates among Anglophone scholars over Darwin's attitude towards humans in the *Origin.* However, translations of his text added stress and even extended the consequences his thinking had for humans. By the mere effect of translation, the presence of the "race" question was amplified exponentially to spread throughout the book. Other textual effects (like Royer's preface) also strengthened the impression that the *Origin* was actually a book about human races.

... in the struggle for life

Considering the "favoured"—and thus, implicitly, those "unfavoured" or "doomed"—races also draws our attention to the last fundamental concept of the title: "*struggle for life*" (or "*struggle for existence*" in the title of Chapter III).

The phrase "struggle for existence" pre-dates Darwin. Conway Zirkle has identified the idea (if not the term itself) in a study where he compiles several possible "forerunners" to "natural selection":[86] Zirkle gives quotations from numerous natural philosophers (such as Empedocles, Lucretius, and Diderot), anthropologists (such as Prichard), and naturalists (such as Naudin), but his interpretations often read more "Darwin" into these early writers than is actually justified, equating natural selection with either mere elimination or else with the expression of necessity. All literature covering "checks to an excessive increase", such as Matthew Hale's 1677 *Primitive Origination of Mankind*, together with depictions of a "*bellum omnium contra omnes* (war of all against all)", from Hobbes and others, are generally grouped under this "struggle for life" heading. It is ironic that the phrase *bellum omnium contra omnes* was picked up by Marx and Engels, who thought Darwin guilty of transferring the idea from society directly into nature— but, as Gillian Beer has shown, Marx reads Darwin as though the latter had brought no change to the meaning of the phrase.[87] As a matter of fact, "struggle for existence" or "for life" is not synonymous with Candolles' "nature at war".

Darwin owes his reference to a "struggle for existence" to his reading of the political economist Thomas Robert Malthus, something he mentioned in several places, from the *Origin* to his *Autobiography*.[88] Malthus' influence is also to be seen in the fact that Wallace too, having read the economist's writings, used the phrase in his 1858 paper delivered at the Linnaean Society.[89] The actual role Malthus played in the development of Darwin's idea has been the source of one of the most enduring debates within the Darwinian industry.[90] Some portray Malthus as a catalyst for an already developed theory, while others argue that

Malthus' principle of populations provided Darwin with a mathematical framework that was crucial to the invention of the theory itself. That this issue has been the centre of so much heated scholarly contention, it is because, through the "struggle of existence", Darwin seems to have incorporated the ruthless competitive ethos of Victorian capitalist ideology and laissez-faire politics into nature itself.[91] In reality, however, Darwin also related the idea to Candolle and Lyell, the latter having been an influence on both Darwin and Wallace.[92]

Another issue has been whether the "struggle for existence" depicts an actual combat or whether Darwin uses it merely as a metaphorical expression, as he himself suggests in the third chapter of the *Origin*.[93] This metaphorical character was not lost on some readers. Petr A. Kropotkin, for example, states that, even if Darwin's expression, "the struggle for existence" evidently refers to a process of "extermination", "it can by no means be understood in its direct sense, but must be taken 'in its metaphoric sense' ".[94]

What happened to the phrase "struggle for existence" when it was communicated through the translation prism? In Italian, the expression "*la lotta per l'esistenza*" is disconcerting on the ear,[95] and "*lutte pour l'existence*" sounds in no way natural in French. Claparède, who introduced the latter, expressed his regret at employing "such a barbarous expression". Clarifying this, he said:

> Strictly speaking, it is the combat beings bring against each other in order to battle over their existence. Expressions like *combat de la vie* (combat of life) or *lutte de l'existence* (struggle of existence) just do not carry this meaning.[96]

But it should be noted that the term is no more natural in English: Theophilus Parsons speaks of it with hesitation: "Therefore there must be a competition, or as [Darwin] phrases it, a 'struggle for life' ". Parsons never employs the expression without the protection of these scare quotes.[97] Studies carried out on the Russian intellectual reception of the *Origin* have shown that the expression *struggle for existence* "was at best imprecise and confusing; at worst, and this was much more common, fallacious and offensive".[98]

To avoid this compound form that irritated French ears, offended Russian scholars, and was ultimately just as disconcerting in English as it was in Italian, Clémence Royer proposed a vibrant and illustrative translation: "*la concurrence vitale*" (the vital competition). But what does this version suggest? Given that it is not a literal translation, what does it add to Darwin's text? Whereas struggle indicates *adversity* or *rivalry*, competition primarily indicates a parallelism: the term seems to indicate several individuals concomitantly engaged in the career of life. Nevertheless, it can be noted that where *struggle* may describe both organisms struggling amongst themselves (dogs who are hungry) as well as sole individuals confronting a hostile environment (the plant in need of water at the edge of the desert), "competition" immediately suggests the idea of obtaining some reward or prize. Both "struggle" and "competition" alike bring a certain emphasis to the "relationships of mutual dependence" that individuals maintain both with each other and with their environment: but where "struggle" may include a form of *conatus*, or a simple tendency to persevere in existence despite adversity, "competition" insists more directly on the very relational nature of survival: only one seed in a million reaches maturity. Where the "struggle *for existence*" includes the relation to inert elements and rare resources (water, nutrition, etc.), *vital competition* reinterprets these as relations of rivalry, "the banquet of life ... ever too cramped to allow for all living creatures to be seated".[99]

Adolphe d'Archiac, a French geologist, proposed to replace Royer's "vital competition" with "balance of the vital forces which give rise to the harmony of nature".[100] But even though Royer did sometimes use the term "*harmonie*" to translate the English "adaptation", this is still a far cry from making "*l'harmonie de la nature*" the central lesson of Darwinism, if by "*harmonie*" one means balance and status quo. Rather, as she put it in the 1862 French title, Darwinism appeared to her to be a theory of "progress". More than just "favoured races"—which would at times be interpreted as perfecting (*Vervol-lkommnung*) instead of in terms of simple advantages (*Begünstigung*)—Royer gives a general hypothesis on the nature of Darwin's philosophy. She speaks of natural selection as a "*pouvoir intelligent*", an angle shored up by some few instances of a "perfecting" vocabulary in the *Origin*, or by the fact that Darwin spoke of the "unerring skill" of natural selection.[101] According to Royer, progress is a conclusion logically deduced from the book itself. French readers of Royer's translation went on to echo this considerable inflection, which they unquestioningly attributed to Darwin himself.[102] Darwin did, however, succeed in having Royer remove the concept of "progress" from the subtitle of the second French edition, although this did not prevent the Darwinian theory being lauded by an editor of *Civiltà cattolica* as "*la teorica del progresso*".[103]

"...or..."

In final analysis, might all these rival interpretations of the nature of the Darwinian theory and of its coherency stem from the interpretation of just one very small word: the *or* which connects both halves of the title? Precisely, the function of this connector can be understood in two ways. Up to this point, we have taken for granted that natural selection is the same as "the preservation of favoured races in the struggle for life" and that, consequently, natural selection is defined as a dividing blade, a process that sorts out the "races". This is how the title was principally understood in the nineteenth century. The second interpretation would establish an equivalence, not between *natural selection* and *preservation of favoured races*, but rather between the entire first part of the title and the entire second part of the title, i.e. between *the Origin of Species by Means of Natural Selection* and *the Preservation of Favoured Races*. This would boil down to saying that species have an ontological or taxonomical status equivalent to that of "races" (i.e. varieties); that the *production of species* (rather than *natural selection*) is nothing other than the preservation of favoured races in the struggle for existence. The second reading is supported by present day philosophers of biology such as Jean Gayon:

> This title can be understood in at least two ways. The first suggests that natural selection involves "the preservation of favoured races in the struggle for life", and thus consists of the selection of races. The second (*which was clearly Darwin's*) suggests that species, which cannot be distinguished by any absolute criterion from "races" (or "varieties"), are the result of a process of modification by "natural selection".[104]

In this reading, difficult terms are deflated, separated from tendentious interpretations: the term "races" is simply equated with "incipient species"; the term "preservation", with its toxic hints of elimination, is not associated with natural selection (a creative, non-eliminative process), but to the overall mechanism of the origination of species. Hence, Gayon's interpretation of the central, balancing "or" between the two halves of Darwin's title undoubtedly solves many difficulties for modern day interpreters. However, in spite of its conceptual clarity and theoretical beauty, this interpretation of the "or" cannot be said to be "*clearly Darwin's own*", since we have seen that Darwin's own comments in his correspondence and in the various phrasings and rewritings of the *Origin*, all lean in the same direction: *preservation* is an equivalent of *selection.*

Concluding this analysis, we see that concentrated into the title alone there is a powerful dose of ambiguity and contention. Nineteenth century readers endlessly wondered whether Darwin had really articulated the right problem, and whether he had attacked it from the right angle. Although some did acknowledge Darwin's virtue in having posed an important question, this was sometimes just another tactic for casting his method into doubt and coaxing him towards other wordings, other answers, other problems. As for the solutions he delivered, criticism of these was, for all intents and purposes, universal. None seem to have grasped that Darwin had transformed and shifted the question of origin from a point-source derivation to the mechanisms enabling the efficient production of species (the production of variations, their accumulation and perpetuation). How this reinterpretation of the term "origin" was accounted for would underpin the status accorded to the doctrine of natural selection *qua* response to the "origin of species" question.

Two points in particular seem to emerge from this. First, natural selection's mode of operation (interpreted as a dividing blade, as elimination) leaves the question of the genuinely creative process (variation) wide open. Second, it is because readers either reject or disregard the meaning of the term "origin" that they speculate that the 1859 work may have approached the problem from the perspective of a secondary action, whereas the solution would have required a shift to a more fundamental level. In either case, it seems that as soon as the origin of species question has been opened, we find ourselves inevitably redirected towards other questions that the former presupposes (e.g. the origin of variations, the origin of prototypes, the origin of life, etc.), domains for which the Darwinian model of natural selection could be neither a relevant nor a sufficient agent.

Most often, Darwin tended to reconcile opposing views rather than trying to decide between them, as can be seen in the case of "origin as mode of origination" and "origin as source". Perhaps part of the difficulty lies in the fact that the documentation on breeding that Darwin leaned on tended to conceive of the origin as a *prototype* and variation as a temporary distancing from this type that nevertheless remained ever at risk of reversion. Despite his best efforts to dissociate himself from this prototype conclusion that the breeders had drawn from their own practices, and despite having criticised them for not seeing that the accumulation of variations from just one type could indeed lead to numerous variants and eventually species, it may still be the case that Darwin never managed to fully free himself from the framework in which these problems were initially set in the sources he drew upon. And this in turn may be the root of the book's most commonly received criticism: that it simply doesn't answer the question posed by its title.

Notes

1 Beer 1983, p. 64.
2 Wollaston 1856, p. 47.
3 Godron 1859.
4 Darwin knew about Godron. See "Books Read" and "Books to be Read" notebook. (1852–1860). CUL-DAR128. Transcribed by Kees Rookmaaker (Darwin Online, http://darwin-online.org.uk/). See also "Historical Sketch", *Origin* 1861, p. xviii. Darwin 1868 also refers to Godron's work.
5 See for instance Archiac 1864, pp. 64–65 and pp. 115–117.
6 Mayr 1982, p. 412.
7 Cronin 1991, p. 430.
8 Gould 1992, p. 54.
9 Sober 2000, p. 146. See also Sober 2011, p. 15.
10 Jones 1993.
11 Dennett 1995, pp. 42–47.
12 Wallace (J.) 1995, p. 4.
13 I pass over the American editions here since they required no translation: the second American edition was composed of the second English edition with a few additions from the third; the third American edition was based on the fifth English edition.
14 Livingstone 2014, p. 4.
15 Rupke 1999, p. 336.
16 *Ibid.*, p. 331.
17 Livingstone 2014, p. 11.
18 *Ibid.*, p. 5.
19 For instance, Elshakry 2008.
20 See especially Conry 1974 and more recently, Prum 2014. On Royer, Harvey 1997 is much more balanced and accurate than Conry's. See also Fraisse 2002.
21 Hoquet 2013.
22 Büchner 1869, p. 27; Bennett 1870, p. 30.
23 Loewenberg 1959, p. 13.
24 Wallace (J) 1995, p. 6.
25 On Bronn, see Gliboff 2007 and 2009.
26 For example, Köstlin 1860, p. 2; Seidlitz 1871, p. 20.
27 Cf. Darwin to Victor Carus, 21 June 1869 (Staatsbibliothek, Berlin).
28 Hartmann 1877, p. 82 and p. 97.

29 Gautier 1880, p. xi. See also Claparède 1861 and Büchner 1869, p. 27.
30 Beer 1983, p. 64.
31 Darwin to Murray, 10 September 1859, CCD 7 331.
32 Cf. Darwin to Lyell, 28 March 1859, CCD 7 270. Darwin to Murray, 10 September 1859, CCD 7 331.
33 Especially Ghiselin 1969, p. 89. Stamos 2007 aims to refute this "standard view" of Darwin's "species nominalism". Beatty 1985 (p. 266) emphasised the strategic aspects of Darwin's reasoning on species, especially the fact that the *Origin* was addressed to naturalists.
34 For the perception of Darwin's nominalism in review of the *Origin*, see Stamos 2007, pp. 12–13.
35 Owen 1860a, p. 532.
36 Bowen 1860, p. 482.
37 Simpson 1860, p. 365.
38 Royer 1862, p. xlix.
39 Braun 1872, p. 28.
40 Quatrefages 1877, p. 26.
41 Sloan 1987.
42 Darwin to Hooker, 24 December 1856, CCD 6 309.
43 *Origin* 1859, p. 485, *Var* 754 (#230).
44 *Origin* 1859, p. 52: *Var* 136 (#74).
45 For a recent comment on this puzzle, see Stamos 2007.
46 Quatrefages 1877, p. 70 and Quatrefages 1870, p. 3.
47 *Origin* 1859, p. 411, *Var* 646 (#2).
48 Valroger 1873, p. 132.
49 Haeckel 1874, p. 91. Cf. also Haeckel 1866, t. II, p. 323ff.
50 Wallace 1889, pp. 8–9.
51 Royer 1862, p. xvi–xvii.
52 *Ibid.*, p. xix.
53 *Origin* 1859, p. 485, *Var* 755 (#231).
54 Darwin to Baden Powell, 18 January 1860, CCD 8 39.
55 Butler 1879, p. 345.
56 Beer 1983, p. 65.
57 See Darwin 1864. Cf. also Di Filippi 1864; Pancaldi 1991.
58 Conry 1974.
59 Flourens 1864, p. 53.
60 *Origin* 1861, p. 85; *Var* 165 (#14.5–9:c.) Royer 1866 picks up this argument, p. xiii and p. 95, note.
61 Beer 1983, p. 69.
62 On all these issues, see Hoquet 2013.
63 See, for instance, Darwin 1873, p. 31, where "*selected for breeding*" was translated as "*choisi pour la reproduction*". See also Darwin 1876a, p. 31.
64 Darwin 1873, p. 4.
65 Darwin 1876a, p. 4.
66 Claparède 1861, p. 534.
67 For a defence of Royer as translator, see Miles 1989.
68 This occurs in one attempt to render some very specific nuances of the German debate. A French translation of Büchner (1869, p. 27) reads: "In Darwin's thought, nature does not cultivate [*züchtet nicht*] as man does, she simply eliminates, she picks out [*wählt aus*], but with neither bias nor design". "(*Dans la pensée de Darwin, la nature n'amende pas* [züchtet nicht] *comme l'homme peut faire, simplement, elle élimine, elle sélige* [wählt aus]*, mais sans parti ni dessein*)".
69 "*Sélire*" would have been a better translation of "*to select*" than "*sélectionner*". In fact, "*sélectionner*" is based on "*sélection*", which is itself derived from the verb

seligere. Thus, the new verb *"sélectionner"* is a lexical monstrosity. Unfortunately, the evolution of French was to confirm Royer's nightmares, since *sélectionner* made it into common language, being now a part of the ordinary lexicon of football players, who, not unlike cattle, dogs, or race horses also have their *"sélectionneur"*. A similar case exists in contemporary French with the term *"solution"*, from the Latin *"solvere"*; the verb associated with this is *"résoudre"*, but people have now begun to use the unseemly *solutionner*, based on the substantive *solution*.

70 Gliboff 2007.
71 Büchner 1869, p. 27 and p. 46.
72 Darwin to Lyell, 30 March 1859, CCD 7 273.
73 Darwin to Bronn, 14 February 1860, CCD 8 83.
74 *Origin* 1859, p. 81 and p. 127, *Var* 271 (#386); Darwin to Bronn, 14 February 1860; to W.H. Harvey, 20–24 September 1860; to Gray, 26 September 1860; to Lyell, 28 September 1860, CCD 8 resp. 83, 371, 389, 397.
75 Quatrefages 1867, p. 137.
76 Notebook D 134 (CDN, p. 375).
77 Quatrefages 1877, p. 68. Similarly, Fee 1864 quotes Restif de la Bretonne's *La Découverte australe*, as a book supporting a thesis similar to Darwin's (namely that, originally, there was only one single animal and one single plant).
78 Lyell to Darwin, 30 September 1860 and Darwin to Lyell, 3 October 1860, CCD 8 400 and 403.
79 Darwin to Lyell, 21 August 1861, CCD 9 238.
80 See below, Chapter 9.
81 *Origin* 1859, resp. p. 23, 15 and 467.
82 *Ibid.*, p. 199, 382, 422.
83 This count was made possible by the different versions available on the website www.darwin-online.org, administered by John Van Whye.
84 Compare, for instance, 1859, p. 28 ("the several kinds, knowing well how true they bred", "the several breeds to which each has attended") with Bronn's translation (1860, p. 34) ("Tauben-Rassen" and "die verschiedenen Rassen, welche ein Jeder von ihnen erzogen") and Royer's rendering of the same passages (1862, p. 51: "les diverses races se reproduisaient", "les diverses races à l'étude desquelles, etc".). The terms from the 1859 version remained unchanged in the 1861 third edition (p. 29) which was the base for Royer's translation.
85 See Winkler's Dutch translation of the *Origin*'s table of contents, especially the first three chapters.
86 Zirkle 1941.
87 Beer 1983, p. 58.
88 *Origin* 1859, Introduction, pp. 4–5. See also *Autobiography*, p. 120.
89 Wallace 1858. See McKinney 1972, pp. 54–55; Petersen 1979, pp. 219–223.
90 On Darwin's "debt" to Malthus, see the synthesis in Mayr 1982, pp. 491–494; Bowler 1989a, pp. 173–174.
91 Gale 1972; Young 1985.
92 *Origin* 1859, p. 62.
93 *Ibid.*
94 Kropotkin 1902, pp. 63–64.
95 Cf. Anon. 1871, p. 295.
96 Claparède 1861, p. 535.
97 Parsons 1860, p. 1.
98 Todes 1989, p. 3.
99 Royer 1880, pp. 734–735.
100 Archiac 1864, p. 71.
101 Royer 1862, pp. 271–272 gives *"pouvoir intelligent"* where the English text speaks

simply of "power". Conry 1974 (pp. 263–264) pulls no punches in countering Royer, guilty of "handicapping the theoretical introduction" of Darwinism into France. But Darwin himself, in the introduction to *Variation* (1868, vol. 1, p. 6) wrote: "For brevity sake I sometimes speak of natural selection as an intelligent power; in the same way as astronomers speak of the attraction of gravity as ruling the movements of the planets".

102 For example, Dally 1868, p. 40; Simon 1865, p. 7.
103 Anon. 1871, p. 297 which leans on expressions from Chapter IV of the *Origin*, but which also recalls that, according to Darwin, the theory of natural selection implies neither development nor progress.
104 Cf. Gayon 1998, p. 62 (emphasis added).

3 "One long argument"?

Darwin-the-Selectionist

What is specifically "Darwinian" about Darwin's theory? This needs to be clarified so that any false accusations of plagiarism or decrying of inconsistent principles can be avoided. The ill-fitting labels it has been dressed up in must be stripped away if we are to be spared both irrelevant praise and misdirected condemnation. Now that we have unpacked the various meanings enfolded in the title, we can risk stepping into the maze of the book itself. Here we will see that the *Origin*, even taking only its first edition, can also be read through a prism that separates several distinct Darwins.

The *Origin*, Darwin claimed, consists of just "one long argument". One 490-page long argument; not quite the page-turner that keeps you on the edge of your seat all the way through. Yet, in spite of its length, the book was an immediate, considerable, and also durable success. No doubt some of this is due to various misunderstandings regarding what the book was, in fact, about. What did Darwin really put in his book? What did his readers actually find in it? The answers to these questions are not necessarily unique and transparent.

However, and Darwin insisted on this, the *Origin* must be read as a whole, from beginning to end. As he wrote to Lyell on 2 September 1859, as he was progressively sending his corrected pages to his editor John Murray: "I hope that you will read all, whether dull (specially latter part of Ch. 2.) or not, for I am convinced there is not a sentence which has no bearing on whole argument".[1] Darwin speaks of the *Origin*, "this whole volume", as "one long argument".[2] The "argument" stretches over all fourteen chapters, giving the book both its coherency and its strength. And yet, modern readers of Darwin have difficulty grasping the unity of the case laid before them; if they survive past the voluminous detailing of domestic pigeons in the first chapter, they run out of steam after Chapter IV and are often tempted to skip straight to the conclusion.

How then should this "one argument" be understood? Are we dealing with just one line of argumentation? A single, meditated thread of reasoning that unfolds throughout the book? Or is it one long battle, where "argument" is taken to mean dispute; a battle fought skirmish by skirmish that has Darwin crossing blades with both special and local theories of miraculous creation come what may?

"When I say 'me'", Darwin explained to Asa Gray, "I mean only *change of species by descent*".[3] Pursuing this hint, the neo-Darwinian tradition answers that the *Origin* is a theory of *descent with modification*. However, this leaves some-thing unsaid: that, *relying on natural selection*, Darwin's theory is different from all other theories of descent or evolution. It is striking that the phrase "descent with modification" oddly omits natural selection—the actual lynchpin of Dar-win's theory—from his legacy. Most importantly, the theory Darwin constructs in the *Origin* stands in opposition to theories of special creations. Its strength resides chiefly in a change of world view (evolution vs. creation, transformation vs. fixity) and in its intention to explain the results of its rivals using different means (natural selection rather than metamorphosis or will). However, following Darwin through the twists and turns of the *Origin*, we encounter several occa-sions where he refers to what he calls "my view". What does he mean by this? In certain passages, "my view" refers to *descent* with modification, in others, to *natural selection*, while in others still it seems to pertain to a looser idea of *evolution* in general.

In this chapter, I show that there are, in fact, two opposing ways of reading the *Origin of Species*. The first focuses on the "long argument". This leads to a unitary view of what Darwin was seeking when he wrote the *Origin*: demon-strating that natural selection is the main means of modification leading to the origination of species. The second focuses on Darwin's reference to what he calls "my view". In this reading, we understand that Darwin fluctuates consider-ably with respect to what he considers to be "his". These two diverging ways of reading the *Origin*—focusing either on the unity of the argument or on the mul-tiplicity of the views—are not mutually exclusive, just as it is simultaneously true that the light is white and that it contains a full spectrum of colours. Both readings originate from the characteristic features of the *Origin* as a book and from the history of its writing. Finally, I consider the strange ways in which Dar-win's famous diagram (the only illustration to be found in the *Origin*) was read at the time. Far from being perceived as a key element of Darwin's selectionist views, it was most of the time neglected, if not criticised for its abstract character.

An abstract

The *Origin of Species*, which brought the theory of "descent" onto the naturalist stage, is an "abstract". Darwin, who had been working since approximately 1837 on the question of the "transmutation of species", had already begun putting together a more complete, more developed treatise on "natural selection" which he never found the time to accomplish. Shaken by a manuscript sent to him by Alfred Russel Wallace, Darwin believed there to be such close proximity between their two theories he promptly gave up on the idea of finishing his own book, despite its being well under way. It is in this sense that the *Origin* is an abstract of Darwin's unfinished manuscript; hastily written and completed within less than a year. Evidence for this "synopsis" character of the *Origin* can be seen in the relative dearth of facts and source references that Darwin put into it (viz. the naturalist literature which he had read and annotated in depth), as well as by

the fact that he uses no footnotes. The *Origin* was written in haste, by a man suf-
fering from weak health. This hasty production can be interpreted in two ways.
On one hand, it is what makes the *Origin*'s argument so concise and pithy: "one
long argument". But, on the other hand, it also accounts for the fact that Darwin
had to spend thirteen more years of his life, from 1859 to 1872, editing the book
and improving on the communication of his meanings: wavering as to the
content of his "view", yet rigidly opposing special creations.

The content of the argument

From its table of contents, the *Origin* is a book composed of fourteen chapters.
This was finally changed in the sixth edition (1872) where a new chapter containing "miscellaneous objections to the theory of natural selection" was introduced after Chapter VI, which raised the total number of chapters to fifteen.
Below is a brief overview of the fourteen chapters from the initial version of
the book.

Between the introduction, declaring that "the origin of species" is "that
mystery of mysteries", and the conclusion to Chapter XIV, whose goal is to
retrace the steps of the "long argument", we can separate three principal movements within the main work. The first movement contains Chapters I to V. It can
be read in two different ways. According to the most classical reading, Chapters
I to IV establish the existence of natural selection. The summary of Chapter IV
captures this general dimension. At the centre of this movement, we find two
crucial chapters: Chapter III, on the struggle for existence, and Chapter IV,
which focuses on the notion of natural selection. One can also read these five
first chapters as being unified by the question of variation. Chapters I and II
describe variation in both its domestic and wild states, Chapter V broaches the
actual laws of variation themselves. Several times over, Darwin underlines the
importance of sex in its relation to variability, most notably in a section of
Chapter IV entitled "Sexual Selection".

Following this is a second movement (Chapters VI to IX) where Darwin
examines various difficulties he has identified. In this section, Darwin's rhetoric
is more defensive, nearly sceptical. Almost nothing is actually established and,
for the most part, he stops short at simply asking the reader not to consider the
presented objections as irrefutable. In this, Darwin does not claim to be capable
of "explaining" the assembled phenomena. He suggests only that the facts themselves are either not entirely certain or else do not constitute insuperable objections to his theory. Chapter VI raises the following issue: had structures been
created "for beauty in the eyes of man", says Darwin, then this would be "absolutely fatal to my theory", to which he adds that, had the structure of any species
been formed "for the exclusive good of another species, it would annihilate my
theory".[4] Similarly, in Chapter VII, the existence of insect neuters is considered
to be "one special difficulty, which at first appeared to me insuperable, and
actually fatal to my whole theory".[5] However, in each case, Darwin judges the
facts in question to be either insufficiently established or simply incapable of

overturning his theory. Each chapter presents a specific type of problem. In Chapter VII, on instincts, the problem is the accumulation of gradual modifications by means of natural selection. The end of Chapter VI gives a perfect example of Darwin's intention to revisit and better explain older theories: "two great laws—Unity of Type, and the Conditions of Existence" are re-interpreted in accordance with the theory of descent. Unity of Type refers to "that fundamental agreement in structure, which we see in organic beings of the same class, and which is quite independent of their habits of life". For Darwin, this structural agreement (attested by comparative anatomy) is easily explained through the simple relation of descent; behind it is simply "the inheritance of former adaptations". The law of the conditions of existence is far more general: it refers to the adaptation of all the parts of an organism to its conditions of life, both organic and inorganic. Darwin states that this fundamental law of adaptation "is fully embraced by the principle of natural selection". Indeed, natural selection, far from limiting itself to observing the organism's "adaptation" to its conditions of existence (being equipped with all that is necessary for its survival), actively produces these adaptations from the very variations of the organism's parts. Thus, where the "great laws" of the creation theories were limited to repeating descriptions of facts, Darwin proposes explanatory mechanisms.[6]

Chapter VII applies the same treatment to theories of instinct. Instinct had often been explained away as "habit", but Darwin distinguishes between the two notions. The old "habit" comparison "gives, I think, a remarkably accurate notion of the frame of mind under which an instinctive action is performed, but not of its origin".[7] Rather, Darwin says that he "can see no difficulty in natural selection preserving and continually accumulating variations of instinct to any extent that may be profitable", just as "modifications of corporeal structure arise from and are increased by use or habit".[8] Chapter VIII, on hybridism, endeavours to blur the frontier between variety and species. This chapter tends to establish that "sterility is not a specially acquired or endowed quality, but is incidental on other acquired differences" (those affecting the reproductive system).[9] Professing ignorance, whereby Darwin does not so much explain a fact as create a niche where some new hypothesis might slide in, is a tactic he especially employs in Chapter IX, on the "imperfection of geological record". What is striking in this chapter is to see that Darwin does not use findings from geology and palaeontology to either prove or shore up his conception (what he calls "my view"). Instead, he seems content with showing that they are not in contradiction with his theses. The problems in question are the existence of intermediate varieties, the length of time necessary for the appearance of a species compared with the historical length of geological periods, and the question of whether or not species appear suddenly in the earth's records (and, if they do, why).

The third movement (beginning with Chapter X) allows Darwin to evoke various disciplines and present how his theory enables their results to be accounted for. In particular, this explains why Chapter X returns to ground covered in Chapter IX: Darwin uses it to turn the argument of imperfection around. On the one hand (Chapter IX), the geological records (because of their lacunar nature) present no definitive contradiction to the theory of descent, but on the other hand (Chapter X), we do find in the fossil record "evidence of the slow and scarcely sensible mutation of specific forms",[10] in exact conformity with what the theory of descent should have us expect. Other sets of facts are called upon in later chapters: geographical distribution (Chapters XI–XII); "mutual affinities of organic beings" (Chapter XIII) attested to by morphology, embryology, and rudimentary organs, to which can be added the classification or "arrangement" of organic individuals and groups.[11] As regards these last facts, Darwin applies an annexation strategy vis-à-vis the work of his predecessors: "community of descent is the hidden bond which naturalists have been unconsciously seeking".[12]

Contrary to other theories in vogue at the time that also explained the form, diversity, and adaptation of living species (special creations, unity of type, conditions of existence, etc.), the *Origin* presented a twofold theory of descent (common descent and descent with modification). Darwin himself underlines this double aspect of descent when, in closing Chapter XIII, he concludes "that the innumerable species, genera, and families of organic beings, with which this world is peopled, *have all descended*, each within its own class or group, from common parents, and have all been modified *in the course of descent*".[13] Chapter XIV offers a recapitulation, underlining, throughout, the current theory's opposition to any theory of special creations.

Species as marked varieties

Throughout his book, Darwin often employed "my view" to indicate his opposition to theories of special creations. Species are not specially created; they originate. And precisely, on the question of "the origin of species", the specificity of the Darwinian theory is this: "our view, that species are only strongly marked varieties with the intermediate gradations lost", i.e. varieties are "incipient species", and: "our view that species of all kinds are only well-marked and permanent varieties".[14] In other words, to give the origin of species is to grasp them "*in statu nascenti*".[15] The introduction to the *Descent of Man* enables these points to be laid out clearly, particularly the opposition of the theory of *descent* to the theory of special creations:

> The sole object of this work is to consider, first, whether man, like every other species, is descended from some pre-existing form; second, the manner of his development; and third, the value of the differences between the so-called races of man.[16]

Three points are put forward: descent, development, and value of differences. This is why Darwin's theory, consisting in the affirmation that species share a "*common descent*", is not in itself a new one.

> The conclusion that man is the co-descendant with other species of some ancient, lower, and extinct form, is not in any degree new. Lamarck long

ago came to this conclusion, which has lately been maintained by several eminent naturalists and philosophers.[17]

To write a treatise on the origin of species is to commit oneself to explaining how species give birth to each other, how one spawns from the other, and how the fact of their common ancestry explains their resemblances. This theory of *descent*, relying on natural selection, is not just opposed to other contemporary theories; its strength resides primarily in its aim to explain the results of its rivals, in particular theories of creation. On numerous occasions, there is clear opposition between Darwin's "view" and the idea of miraculous divine intervention into the order of nature. In the science of his time, theories about centres of creation, or *foci*, multiplied acts of creation in space while "separate" creations multiplied them in time. The former kind involved geographical distribution: rather than a single centre of creation (Eden), they posit a polycentric hypothesis, whereby each centre of creation constitutes a point of propagation. The latter were rooted in palaeontology and stratigraphy, in the observation that species have disappeared and appeared successively in time.

Darwin set himself against both types of multiple creation theories and was well-read on the arguments of those who did likewise, in particular Charles Lyell, who developed a "steady-state" vision of the history of the earth.[18] Lyell's *Principles of Geology* became a model for Darwin because it provided a good example, both reputable and recognised by contemporaries and compatriots, of what science should do.[19] The theory of *descent with modification* was Darwin's attempt to achieve for the origin of species what Lyell had achieved for geology: a bold project when one considers that Lyell had openly and expansively opposed the idea of species transformation. Darwin read Lyell's work attentively and several passages of the *Origin* bear witness to this. In order to win Lyell over, to facilitate his adoption of the ideas presented in the *Origin*, Darwin insisted upon the continuous but gradual nature of transformations and the slowness of evolutionary processes. Specifically, he claimed that his theory would undoubtedly meet with the same objections Lyell's had, because both banished sudden changes from their theories (for Lyell, the diluvian hypothesis, for Darwin, continuous creation or sudden changes in structure), and also because both relied on their own "trifling and insignificant cause":[20] tide action for the one, the accumulation of infinitesimal hereditary modifications for the other.

The structure of the argument: *vera causa* and consilience

The question of when Darwin really achieved a consistent argument is still a confused one. Ernst Mayr (1982) and David Roger Oldroyd (1984) both drew up recapitulations of the various views. All authors, Mayr claimed, "are in agreement ... that the theory evolved slowly and piecemeal".[21]

However, Mayr thinks that "indeed, even in his later writings Darwin is often inconsistent when referring to selection and he makes statements occasionally that are incompatible with other statements made almost simultaneously".[22] In spite of these inconsistencies, the Darwin Industry has proved a rich source of clever attempts to reconstruct the structure of the *Origin* so as to sketch a definite outline of what the book's 490 pages are really about. According to these attempts, the *Origin* is "one long argument" in favour of evolution by natural selection.[23] To support this hypothesis, scholars have brought epistemological models to the fore that were in use in Darwin's times: *vera causa* and *consilience*.

First, the structure of the Darwinian argument can be analysed using the "*vera causa*" theme. From this viewpoint, the goal of the *Origin* would be to show that natural selection is a *vera causa*. The *vera causa* concept is of Newtonian origin, possibly communicated to Darwin through John Herschel.[24] When determining the cause of a phenomenon, Herschel advanced a method of analogical reasoning that bases itself upon what is already known. We can be certain of having found the *vera causa*,

> [if] the analogy of two phenomena be very close and striking, while, at the same time, the cause of one is very obvious, [then] it becomes scarcely possible to refuse to admit the action of an analogous cause in the other, though not so obvious in itself.[25]

According to Jonathan Hodge's classical interpretation, the whole structure of the "long argument" in the *Origin of Species* consists of presenting natural selection as a *vera causa*.[26] Such an interpretation would explain the structure of the first four chapters as well as the comparison Darwin draws between domestic variation and variation in the wild: if, in the first case, the production of different breeds is the effect of selection, one is to suppose that an analogous selection occurs in the second case. It then remains for him to show that such a selection does exist (the very *raison-d'être* of the "struggle for existence" concept) and then show how the production of breeds, over the course of generations, is equivalent to the production of species and groups of increasing taxonomical variance.

In accordance with this interpretation, Hodge has proposed that the *Origin* be redeployed as a three-step argument establishing the following properties of natural selection:

- its *existence* in Chapters I to III: it continues to operate in nature, there are conditions of variation in both domestic and in wild species; there is also a struggle for existence which intervenes in much the same way as the different reproductive rates of species;
- its *adequacy* in Chapter IV: it acts, far more efficiently than "man-made" selection, towards the formation of species and their adaptive diversification over time into genera, families, etc.;
- its *responsibility* in the remaining chapters: probably it was the main agent in the production of species still roaming the earth today, just as it was for those now preserved only in the buried rocks of prehistory.

This interpretation sheds excellent light on the structure of the opening movement, clarifying in particular Darwin's decision to begin with an analogy between natural selection and artificial selection.

Another comparable concept, borrowed from William Whewell and brought to general attention by Michael Ruse, enables the importance of the *last* chapters to be accounted for: this concept is *consilience*.[27] Consilience (literally "jumping together") occurs when many independent sources converge in pinning down some particular phenomenon. Thus, consilience of inductions refers to an argument strategy that consists of reinforcing an analogy through the convergence of disparate results coming from various sources. Its emphasis is on the predictive aspect of theory. In fact, the *Origin* closes with a tour-de-force that enables the derivation of various domains of phenomena, apparently distant, all from one and the same body of principles or axioms. This is precisely what Newton had so masterfully accomplished in proposing, in the force of attraction, a concept that causally unified various sets of facts (planetary movement, the tides, falling apples, etc.). As such, consilience of inductions would be a strong indicator that the *vera causa* was to be found at the locus where a whole set of analogies intersected. This consilience model draws our attention to the third movement of chapters identified above: Chapters X (fossil record), XI–XII (geographical distribution), and XIII (affinities, morphology, embryology, systematics). Together, these diverse groups of facts form a corroborating set that qualifies natural selection as a plausible analogy.

Vera causa and consilience are useful tools for understanding from whence the strength of Darwin's theory comes and why it is not merely another evolutionary hypothesis. Why was Darwin's *Origin* not just lumped in with the fantastical evolutionary speculations of Benoît de Maillet, Jean-Baptiste Robinet, or even Darwin's very own grandfather Erasmus Darwin? For the American botanist Asa Gray, the doctrine of the *Origin* draws its strength from its ability "to show the general conformity of the whole body of facts to such assumption, and also to adduce instances explicable by it and inexplicable by the received view". Its whole seductive, persuasive power comes from "its competency to harmonize all the facts, even though the cause of the assumed variation remain as occult as that of the transformation of tadpoles into frogs, or that of *Coryne* into *Sarzia*".[28] Darwin admits that many difficulties can be raised in opposition to his hypothesis, but natural selection (shored up by the domestic productions analogy, the existence of the struggle for existence, and the variability of organic beings) seems to him to be just as well founded a hypothesis as the aether wave theory.[29] In fact, Darwin's strategy on both fronts is a defensive one: acknowledging the hypothesis status of his work while at the same time, off the back of a solid analogy, asserting its status as a *vera causa*.

The methodological approach of the *Origin* has proved very efficient in responding to classical attacks against Darwin's work, such as Gertrude Himmelfarb's caricatural account, according to which Darwin "gives the appearance of an amateur, an amateur even for his own day".[30] However, I think the methodological approach of the *Origin* has three main faults.

First, Darwin's methodological argument was not recognised and approved by his "patrons" in scientific methodology. As Michael Ruse himself clearly puts it: "In England the reception of the *Origin* by the philosophers of Darwin's day ranged from outright hostility to lukewarm acceptance mixed with misunderstanding".[31] These considerations, aligning Darwin with the epistemological models of his time, are difficult to follow precisely because the *Origin* has at times been faulted not only for erring in the arguments it presents but also for its methodological shortfalls. Herschel, despite acknowledging that the introduction of a new species was a natural process and not a miraculous creation, cruelly declared Darwin's theory to be nothing more than "the law of *higgledy-piggledy*" and judged it inadmissible, like a "*Laputan method for composing books*".[32]

Given Bacon and Newton's status as the Grand Lord Protectors of natural philosophy in Darwin's Britain, the latter's method was immediately held up to the question of *induction*. Adam Sedgwick, on 24 November 1859, confessed his own deep disappointment to Darwin directly: "You have *deserted*—after a start in that tram-road of all solid physical truth—the true method of induction".[33] John Stuart Mill seems more merciful in his *System of Logic* when he states that "natural selection" is "not only a vera causa, but one proved to be capable of producing effects of the same kind with those the hypothesis ascribes to it". But then he adds:

> it is unreasonable to accuse Mr. Darwin (as has been done) of violating the rules of Induction. The rules of Induction are concerned with the conditions of Proof. Mr. Darwin has never pretended that his doctrine was proved. He was not bound by the rules of Induction, but by those of Hypothesis.[34]

As Michael Ruse put it, "Mill was offering half a cake at the most".[35] In reality, this amounts to stating that Darwinism is not "proved" and that it is thus not a "theory", in the proper sense, but, at most, a "legitimate hypothesis" not bound to the rules of induction.

The second uneasiness I have with a full, consistent acceptance of the role played by methodological principles in Darwin's *Origin* comes from the text of the *Origin* itself. In it, Darwin makes no reference to *consilience* but does evoke "*vera causa*" on three separate occasions: in Chapters V, XI, and XIV. Strikingly enough, the *Origin* never applies the expression directly to natural selection but rather to theories of modification in general.

In Chapter V, "community of descent" is presented as the *vera causa* underpinning the similarity of three plants, countering "the ordinary view of each species having been independently created".[36]

In Chapter XI, countering the hypothesis of "single and multiple centres of creation", Darwin supposes a unique cradle of creation from whence species must have migrated (even if, given the immensity of the natural barriers involved, this migration may be difficult to conceive of). In multiplying the acts of creation, one necessarily "rejects the *vera causa* of ordinary generation with subsequent migration, and calls in the agency of a miracle".[37]

In Chapter XIV, countering expressions that explain nothing and serve merely to mask our ignorance (plan of creation, unity of design), Darwin condemns those naturalists who admit that certain species were produced by secondary causes (variability) yet simultaneously insist that other genuine species were specially created: "They admit variation as a *vera causa* in one case, they arbitrarily reject it in another, without assigning any distinction in the two cases".[38]

In all three instances, the *vera causa* asserts the action of secondary (natural) causes in opposition to miraculous interventions (special creations).

The third principal objection I have with the methodological approaches to the *Origin* is what they leave unexplained. However central and illuminating they may be, these borrowed concepts from the history of scientific method (*vera causa*, consilience) come with their own inconveniences: they skip over the entire central movement of the book. Notably, they struggle to accommodate Chapter V on the laws of variation.[39] Few mention the fifth chapter as belonging to the core of the *Origin*. Vorzimmer may claim that "the core of the *Origin* lies in its first five chapters", but he describes them as "those in which Darwin described the conditions and the process of natural selection", that is, he disregards the specific subject matter of the fifth chapter ("Laws of Variation").[40] Similarly, Mayr claims that "Darwin's evolutionary model" "can be reconstructed from the first five chapters of the *Origin*", and yet Chapter V plays no actual role in his reconstruction of the argument.[41] Most scholars readily admit the Darwinian formulation of "one long argument" (and therefore its unity as well), rarely asking what the argument actually involves, what the veritable object of Darwinian theory is, or, as the case may be, against what the argument (i.e. dispute) is levelled. At the same time, there are plenty of papers that treat the "long argument" as a puzzle and instead cast light on the plurality of Darwin's arguments: one argument plainly deals with "natural selection", another has to do with "common ancestry" or "the tree of life".[42] Regardless of how many theories were actually supported by Darwin's *Origin*, when it comes to exposing the structure of the book, scholars tend to reduce Darwin's answer to the puzzle of the origin of species to just one dimension: the theory of natural selection. So if, in fact, the *Origin* is *not* just a book on natural selection, then what is it actually about?

"When I say me…"

Darwin was always quite loose in his use of the phrase "my theory". Having analysed Darwin's transmutation notebooks, Howard Gruber warned readers that they

> should bear in mind that Darwin uses the phrase "my theory" liberally throughout the notebooks, to refer to whatever idea happens to have caught his enthusiasm at the moment, especially when he is thinking of his ideas in relation to those of others.[43]

Hence when Darwin refers to "my view", he does not necessarily mean "natural selection". Chapter XIV, for instance, often refers to the themes of variation, generation, mutability, and modification of species. The theme of natural selection is somewhat down-played, particularly at the end of the book where it is shown how "these elaborately constructed forms, so different from each other, and dependent on each other in so complex a manner, have all been produced by laws acting around us". It is striking to observe how Darwin is content to situate natural selection in a far more general list of biological laws or processes, not even giving it pride of place.[44]

Most notably, all throughout the *Origin*, natural selection is always conspicuously accompanied by the laws of variation. As Chapter IV recalls, the circumstances favouring natural selection are those "giving a better chance of profitable variations occurring; and unless profitable variations do occur, natural selection can do nothing".[45] Chapter X also concludes, from the findings of palaeontology, that "species have been produced by ordinary generation: old forms having been supplanted by new and improved forms of life, produced by the laws of variation still acting round us, and preserved by Natural Selection".[46] Nor is this sentence an addition from some later edition; it is there, in plain sight, right from the 1859 first edition and featured in a particularly prominent place as well. Here, Darwin is simply asserting the following point: the only function of natural selection is to preserve and it falls to generation and the laws of variation to actually "produce" new forms. This same insistence on the two factors of generation and the laws of variation is also found slipped into the conclusion to Chapter XII, where Darwin states, "On my theory these several relations throughout time and space are intelligible"; the forms of life

> within each class have been connected by the same bond of ordinary generation; and the more nearly any two forms are related in blood, the nearer they will generally stand to each other in time and space; in both cases the laws of variation have been the same, and modifications have been accumulated by the same power of natural selection.[47]

The beginning of Chapter XIII also makes joint and equal reference to "our second and fourth chapters, on Variation and on Natural Selection" when indicating which species vary the most and how varieties, or "incipient species", "become converted, as I believe, into new and distinct species".[48] The beginning of Chapter XIV also reveals the increasing space accorded to the laws of variation. Although Darwin openly confesses his acceptance that "many and grave objections may be advanced against the theory of descent with modification through natural selection", he goes on to clarify the point thus; "descent with modification through *variation* and natural selection".[49]

As a result of Darwin's wavering between both selection and variation and between various laws of the organic world, his readers have been especially interested to know what he saw as truly "his". First-person pronouns are heavily present in the beginning of the book. Darwin writes to J.D. Hooker:

Here is a good joke: H. [C.] Watson (who I fancy and hope is going to review new edit[ion] of *Origin*) says that in first 4 paragraphs of the Introduction, the words "I" "me" "my" occur 43 times! I was dimly conscious of this accursed fact. He says it can be explained phrenologically which I suppose civilly means that I am the most egotistically self-sufficient man alive,—perhaps so.[50]

But what does this "my" stand for apart, of course, from Darwin's own over-inflated ego? To Baden Powell, who complained that Darwin could have cited him and acknowledged his theoretical precedence, Darwin replied:

> No educated person, not even the most ignorant, could suppose that I meant to arrogate to myself the origination of the doctrine that species had not been independently created. The only novelty in my work is the attempt to explain *how* species become modified, and to a certain extent how the theory of descent explains certain large classes of facts; and in these respects I received no assistance from my predecessors.[51]

One way to appease these "forerunners" was for Darwin to introduce the "Historical Sketch" in the third edition of the *Origin*.[52]

But Darwin had other ways of complying with his contemporaries' complaint that what the *Origin* advanced was maybe not so new, was maybe just the old, well-worn theory of the transformation of species. Samuel Butler, who we already saw embroiled in the New Zealand debates involving the various versions of the *Origin*, was most eager to track down Darwin's "numerous, successive, slight alterations", especially those that might illuminate what Darwin himself perceived as and included in "my view" or "my theory". What did Darwin actually claim as his own? Butler insightfully documented what he interpreted as signs of panic, or what he calls a true "stampede of my's", in the fifth edition (1869) of the *Origin*. In one chapter of his *Luck or Cunning*, mysteriously entitled "The Excised 'My's'", Butler calculates that Darwin excised no fewer than thirty of the initial forty-five occurrences of "my" from the first edition. Finally, "of the fourteen my's that were left in 1869, five more were cut out in 1872, and nine only were allowed eventually to remain".[53] For Butler, the excision of "my's" plainly signifies that "complaint had early reached Mr. Darwin that the difference between himself and his predecessors was unsubstantial and hard to grasp". By 1869, it was clear enough to most that Darwin had not paid due tribute to Lamarck and that he could not claim the whole of evolutionary theory as his own. In fact, this "stampede of 1869" may well have been occasioned by the publication in Germany of Haeckel's *History of Creation* (1868), in which Lamarck is granted a prominent role. For Butler, readers of the *Origin* necessarily feel "the hesitating feeble gait of one who fears a pitfall at every step". This is why comparing the successive editions of the *Origin* was a duty for any reader of Darwin—or at least so Butler thought.[54]

Samuel Butler, meticulously, almost perversely, and with evident perspicuity, counted the "my's" from one edition to the other. And he noticed a fact that is especially relevant to our investigation into the names and meanings of Darwin's theory. Darwin, it is now clear, often speaks, at least in the first versions of the *Origin*, of "my theory" and then shortly afterwards of "descent with modification" in a way that suggests (but does not explicitly *state*) that the two expressions refer to the same thing. In fact, Butler notes, "I only found one place where Mr Darwin pinned himself down beyond possibility of retreat, however ignominious, by using the words *my theory of descent with modification*". This passage is on page 381 of the first edition of the *Origin of Species*. "Darwin", Butler mercilessly pursues,

> only used this direct categorical form of claim in one place; and even here, after it had stood through three editions, two of which had been largely altered, he could stand it no longer, and altered the "my" into "the" in 1866, with the fourth edition of the *Origin of Species*. This was the only one of the original forty-five my's that was cut out before the appearance of the fifth edition in 1869, and its excision throws curious light upon the working of Mr Darwin's mind. The selection of the most categorical my out of the whole forty-five, shows that Mr Darwin knew all about his my's, and, while seeing reason to remove this, held that the others might very well stand.

Darwin even left "*On my view of descent with modification*" on page 454 of the first edition. In 1866, Darwin excised "the most technically categorical" of his "my's": this first revision betrays the deep uneasiness of mind that would soon lead to the "stampede" of 1869.

"By no means an easy book to read"

To better grasp the extent to which the book consists of a hodgepodge of various layers of views, rather than one single argument, we can turn to certain keen supporters of Darwin and see how they perceived his text after reading it. One would imagine that they, more than anyone, must have penetrated to what the book's "true" core is said to be, i.e. natural selection.

The first example I will take is T.H. Huxley. Known as "Darwin's bulldog", Huxley is the Darwinian par excellence. As early as April 1860, Huxley remarked that, perhaps more than any other book, the *Origin* calls for a clarifying treatment: "notwithstanding its great deserts, and indeed partly on account of them, the *Origin of Species* is by no means an easy book to read—if by reading is implied the full comprehension of an author's meaning".[55]

One cannot express in a clearer way the fact that readers of the *Origin* dug into it with gusto, that reviews of the *Origin* had exploded, but that still it was legitimate to wonder whether all this had really led to any extra clarity, and even whether each new review had not already become superfluous. What was Huxley expressing when he spoke of the book's difficulty? That the mass of reviews distributed had done nothing to shed any light on its central question; that, quite to the contrary, they had covered up and contributed to obscuring the actual issues involved. In a word, that the true work of clarification remained to be done. And yet, it was in a review that these very reserves concerning reviews were expressed.

More than thirty years later, in 1888, a mature Huxley expressed similar difficulties with exposing the general argument of Darwin's book. He wrote the following to Hooker:

> I have been trying to set out the argument of the *Origin of Species*, and reading the book for the *n*th time for that purpose. It is one of the hardest books to understand thoroughly that I know of, and I suppose that is why even people like Romanes get [it] so hopelessly wrong.[56]

This letter is exemplary as, in it, Huxley admits to his own difficulties in reading the *Origin*, all the while explicitly criticising those of George Romanes. In other words, he lifts the veil on the possible errors in his own reading in order to explain (and superficially excuse) what he denounces as the flawed reading of his colleague.

Romanes, as we shall see, questioned the relationship between natural selection and the origin of species. Does this give him licence to call himself a "Darwinian"?[57] To answer that question would be to solve the riddle of the place natural selection holds within the Darwinian system. Having abandoned natural selection, can one still claim to be a follower of Darwin? Huxley wished to exclude Romanes from Darwin's legacy. But is it a certainty that, in terms of Darwinian orthodoxy, Romanes matters less than Huxley? While Huxley certainly holds the reputation for being the most devoted and zealous of all Darwinians, it is nevertheless no longer fully clear whether he truly understood the scope of Darwinian theory and, from the sidelines, Romanes' position has now found its own defenders, even in recent times.[58] We will come back to both Huxley's and Romanes' stances on Darwin in later chapters. For now, it suffices to observe that even Huxley had difficulties exposing the book's general argument.

In a later attempt to recapitulate Darwin's theory (1880), Huxley would take up one of Darwin's own formulas for defining the doctrine of the *Origin*, the closing lines to Chapter XIII where Darwin states "that the innumerable species, genera, and families *have all descended*, each within its own class or group, from common parents, and have all been modified in the course of descent".[59] What is striking here, is that such a characterisation of the *Origin*'s argument makes no mention of natural selection and attributes no explicit role to it: as though natural selection was a piece of ornamentation that the *Origin* could easily dispense with.

The second example I will take is Ernst Mayr. Upon rereading Darwin in the neo-Darwinian synthesis framework of the 1950s, Ernst Mayr was surprised to discover that when he spoke of "my theory", Darwin was often referring to evolutionary ideas "in general" and to the theory of natural selection with far less frequency: in fact, on only three occasions:

> Considering that evolutionary ideas, however vague, were widespread in the middle of the century, it surely appears naïve for Darwin to refer to the concept of evolution no less than ten times as "my theory" (pp. 161, 173, 179, 184, 189, 206, 281, 314, 341, and 454), while his own theory of evolution by natural selection is designated as "my theory" only three times (pp. 199, 201 and 242).[60]

Mayr saw this as a sign of "naïvety", as though Darwin were blind to what his own contribution actually was. Mayr's concern is understandable: if the Darwinian theory is not the theory of natural selection, but only one among many theories of *descent*, then it would find itself engulfed in a vast glut of other such theories! Worse still, it may be that Darwin is just a follower of Lamarck and the other evolutionists.[61]

Mayr also noted that most of the literature on the impact of Darwinism paid insufficient attention to the variety of theories actually attributed to Darwin and how these continuously changed over time: sometimes evolution as such, sometimes man's descent from the apes or natural selection. Mayr later clarified his views, explaining that Darwin had "five theories" and urging readers not to approach the "five strands of Darwin's thought" in the same way, since these "do not constitute an indivisible whole, as was made clear by the fact that so many evolutionists accepted some of [them] but rejected others".[62] Those five theories are: evolution as such, common descent (evolution from common ancestors), gradualness of evolution, multiplication of species, and natural selection.[63]

At various levels, each of these five theories have represented important ways of characterising the Darwinian theory. They all compete to be the best name for designating the ideas presented in the *Origin*. I will focus on the two major names from this list: a theory of "descent with modification"; and a theory that effects transformations by means of "*natural selection*". What are the merits of these various designations?

Merging *Descent* with *Origin*

Ernst Mayr had identified five different theories in Darwin's *Origin*, the second of which is common descent.[64] A first candidate to designate Darwin's theory is that species are "*allied by descent*".[65] But how should this phrase be understood? At least two interpretations are possible: *by descent* can mean "over the course of descent", i.e. all along the genealogical line that runs from the common ancestor to present forms. But a second interpretation suggests that the change of species occurs "by means" of some specific action called "descent", the fact of one individual being the "descendant" of another. In the first case, we turn to the depths of time; in the second, to a mechanism still at work in nature.

There are also two main ways the word "descent" itself can be employed. On the one hand, when Darwin speaks of a "community of descent", of "common descent", or of "propinquity of descent", he points into the past, indicating a "progenitor", a common ancestor, a "community of origin", the foundation of membership to a shared line, i.e. the resemblance between organic creatures. This idea allows him, in Chapter XIII, to account for the natural arrangement of species. But, on the other hand, Darwin evokes the "long course of descent", the modifications that occur at every "step of descent", "successive period of descent", or "long [line] of descent". These passages use descent in the forward-looking sense. Here, descent points towards the descendants, to generation in its double sense: the production of new individuals and the relationship of succession they occupy with their progenitors.

The term "descent" has proven troublesome for translators. In French, the word "*descendance*" has been traditionally used, although it has several drawbacks: it lends itself to much lampooning (humans "*descending*" from monkeys just as monkeys "*descend*" from the tree, a play on words possible in English also but which was famously caricatured for the front cover of the French satirical magazine *La Petite Lune*, circa. 1878). Worse still, it presents some genuine conceptual limits. The French "*descendance*" primarily indicates posterity, whereas the English "descent" also indicates ancestry. Patrick Tort proposed translating "descent" by "*filiation*". To establish the "descent of man", for example, would be to establish the "*filiation de l'homme*", to set down a genealogy, to identify ancestors.[66] This interpretation has unquestionable advantages, particularly because it underlines the systematic and palaeontological significance of descent. A French palaeontologist, Paul Gervais, wrote the following in 1859, just before the publication of the *Origin*:

> The *filiation* of the animal species, through time, is only an apparent filiation, or rather it is a succession of specific terms, in many cases at least, and could not be considered to be a genealogical filiation in the manner of the individuals of some same bloodline.

In this, he was rejecting the idea of a genuine "*filiation*", preferring instead to speak of "*seriation*" to denote "the obvious rapports we observe between the specific forms that represent and seem to perpetuate the same natural group over several successive epochs of the same geological period or over different epochs".[67] Darwin, on the other hand, maintained the idea of genuine "*filiation*".

Darwin's theory is a theory of both common ancestry *and* of modified descent. It identifies "*common descent*" with "*the theory of descent with modification through natural selection*":[68] if the first part of this phrase is opposed to all theories of special creations, the latter specification (*through natural selection*) put Darwin's theory at odds with all other theories of transformation or evolution. Considered as a general hypothesis, *descent* is a more economical explanation, simpler and more efficient than *miracles* and other specific acts of creation. It is a "*vera causa*" based on strong analogies, enabling the explanation of numerous phenomena, but it tends to be mixed up with *development*.

Besides, *descent* must not be exclusively understood as the search for a common ancestor; it is also to be taken in the sense of new species production. In fact, Darwin often employs the terms *descent* and *origin* as equivalents: to give the *origin* of species is, simply, to say that species are the modified *descendants* of other species.[69] The theory of *descent* is a theory of origin, in its double sense of *source* (the relation of progenitors to descendants) and *efficient cause* (the producing mechanism).

Making a metaphor of natural selection

Second in line with a claim to designating Darwin's theory is the term "natural selection". With this term, Darwin points to a certain rational, causo-mechanical explanation of the origin of species. In this perspective, Darwin could not have been the forefather of theories of evolution or descent, though it was he who brought natural selection to the table. In contrast to the *Abstammungslehren* or *Descendenztheorien*, where forms supervene from each other spontaneously in a process often described as going from the simple to the complex, the specificity of Darwin's theory is that it is a *Züchtungslehre* or *Selectionstheorie*.[70] This distinction between *organic evolution* (i.e. the theory of descent in general) and *natural selection* is also a key point for Kellogg's definition of "Darwinism".[71] For Kellogg, organic evolution is beyond the reach of doubt, while "the natural selection theory as an all-sufficient explanation of adaptation and species-forming" has a weakness in its base: the origin of variation.[72]

What is the concept of natural selection that comes out in Darwin's writings? In its simplified form, natural selection can be presented as follows: in each generation of any animal or plant species, there exists great disparity between one individual and the next. These differences arise randomly and, for the most part, are transmissible to the descendants. Furthermore, more individuals are born than the conditions of existence can sustain. From this situation results a struggle for existence in the course of which characteristics that are "useful to each being's own welfare" will be "advantageous to them". And so, "assuredly individuals thus characterised will have the best chance of being preserved in the struggle for life.... This principle of preservation, I have called, for the sake of brevity, Natural Selection".[73] Should this process of selecting the best equipped individuals from within a single species stretch out over a long period of time, Darwin's belief is that a group of organisms will result that is so different to its ancestors that one could rightfully consider its members as belonging to a new species.

The accomplishment of the *Origin* is therefore twofold: on the one hand, it establishes the process of natural selection, and on the other hand, in keeping with a title that announces the origin of species "*by means of* natural selection", Darwin frames it as the mechanism underpinning the *origination (Entstehung)* of species. The first aspect presents a mechanism of remarkable simplicity, but it was the second point—i.e. the actual causal or explanatory power of natural selection—that was to open the floodgates of criticism and praise. In reality, the concept of natural selection hides many difficulties, something the historian of science can get an idea of by consulting either Darwin's notebooks or the two unpublished manuscripts (from 1842 and 1844) that outline his theory. In them, we find Darwin wondering how variation occurs; whether it is determined that it occurs in response to environmental solicitation, or, in the case that it may be spontaneously produced by the body, then according to what modalities, within which limits, etc. While the existence of natural selection was generally granted, criticism focused on its effectivity, which seemed to be governed by the amplitude of the variations upon which it acted.

Darwin had no means by which to directly observe the relationship between natural selection and the origin of species. This point was underlined by Thomas Henry Huxley as early as 1860 and, as Jean Gayon has shown, the objection was not destined to disappear any time soon.[74] The mechanism itself, however, seems to be unobservable. In a famous passage of the *Origin*, Darwin declares,

> It may be said that natural selection is daily and hourly scrutinising, throughout the world, every variation, even the slightest; rejecting that which is bad, preserving and adding up all that is good; silently and insensibly working, whenever and wherever opportunity offers.

From the second edition on, this was transformed into the slightly more specific, "It may metaphorically be said".[75] If natural selection is only "metaphorical", this means, first and foremost, that it cannot be taken literally; there is no intelligent and conscious entity in nature that "selects".[76] The term "metaphorical" is also testament to the actual origin of the concept itself. The fact that it was borrowed from another field—breeding practices—and translated into the theory of the evolution of species. The importance of the metaphor resides, for Darwin, in the fact that "it brings into connection the production of domestic races by man's power of selection, and the natural preservation of varieties and species in a state of nature".[77]

The idea of natural selection is often coupled with the idea of the "*struggle for life*", adapted from Malthus's "struggle for existence". Some have declared that Darwin was beaten to the post on both scores. Historians have searched high and low to find the "precursors of natural selection", and Darwin himself recognised that the "struggle for life" had in some way already been discerned by Charles Lyell and Augustin Pyramus de Candolle.[78] Like "natural selection", Darwin says of the "struggle for existence" that it too is to be taken "in a large and metaphorical sense". In reality, there is not necessarily any "struggle" at all, simply "dependence of one being on another".[79] This metaphorical dimension of Darwinian concepts has been diagnosed as a reflection of Darwin's own Victorian environment: a "*nature, red in tooth and claw*", as Tennyson put it, to which Darwin's theory was simply an ideological mirror.[80]

Even though these important concepts give their titles to Chapters III and IV of the *Origin*, Darwin never once used them as synonyms that would encapsulate the full essence of his theory. Quite to the contrary, in his private correspondence Darwin endlessly bemoans the wording *natural selection*, something he expressed publicly in 1868: "The term 'natural selection' is in some respects a bad one, as it seems to imply conscious choice".[81] For Darwin, it is obviously out of the question that nature could "select" consciously, certainly no more than the breeders, who often *unconsciously* selected domestic species.[82]

The concept holds a complex, analogical relation to artificial selection, one which has been widely discussed in the Darwinian literature. Notably, comparison of the two kinds of selection (artificial and natural) creates the impression that natural selection, just like its artificial counterpart, requires a "selector". Therefore, positing an "isomorphism of the two selections" was bound to feel the brunt of abundant objection: doesn't it expose Darwinism to a kind of latent teleology? Darwin speaks of "selection" in much the same way that chemists speak of "affinities", yet nobody stands up to accuse chemists of attributing subjectivity to molecules. This defence notwithstanding, it was discontent with the term "*natural selection*" that led Darwin, in the fifth edition of his work and upon a suggestion from Wallace, to introduce "the expression often used by Mr. Herbert Spencer of the Survival of the Fittest".[83] However, this newer expression brought with it its own set of problems and, what's more, is also an ill-fitting name to represent the Darwinian theory.

When asked to prioritise which of the two terms, Evolution or Natural Selection, mattered most to him, Darwin answered in a letter to Asa Gray: "Personally, of course, I care much about Natural Selection; but that seems to me utterly unimportant compared to the question of *Creation* or *Modification*".[84]

Imaginary genealogies

Another disturbing issue related to the meaning of Darwin's theory and of the role of selection in the *Origin*, has to do with Darwin's famous diagram. Darwin's diagram is now a central feature of the understanding of Darwin's theory. It is even given as epitomising the essence of Darwin's mechanism.[85] However, when the book first appeared, almost nobody commented on the diagram. The only remarks to be found concerning it revolved around its excessive abstraction: it was compared to the trees established by linguists, but only with a view to lamenting the fact that Darwin clung to an empty set of dotted branching lines whereas linguists provided authentic twigs tracing genealogies. More surprising still is that Darwin never bothered to correct his readers: he never complained that the figure was not commented on and seemed quite indifferent to the incomprehension surrounding what he called his "queer diagram.[86]" One might almost think he was indifferent to the outcome of the specific point which, today, is universally recognised as the very heart of his theory. How should this situation be understood?

What is extremely striking in the diagram is the fact that it illustrates a genealogical *mechanism* rather than, strictly speaking, a genealogy. The diagram is striking for the fact that no form of application accompanies it. Darwin makes no attempt at depicting a genealogical lineage anywhere in his book. This decision not to directly represent lines of descent was made long before, ever since his reading of Robert E. Chambers' *Vestiges of Creation*: "I will not specify any genealogies—much too little known at present".[87] This constant reference haunts Darwin's thought: in publishing the *Origin*, he must not fall victim to the mistakes inherent to *Vestiges*.

Simply put, in the place and stead of applications, Darwin often elaborates imaginary examples.[88] In a passage from Chapter VI (already mentioned in the case of the New Zealand debate), he relates that a black bear was seen "swimming for hours with widely open mouth, thus catching, like a whale, insects in the water".[89] Here, Darwin is not positively asserting anything: he limits himself to merely suggesting the power of natural selection by advancing that this mechanism could even (under certain conditions) transform a bear into something as deviant and monstrous as a whale. However, taking to the task a little too precipitously, a reader on the lookout for genealogical applications of Darwinian theory could be led to believe that the object of this quest had finally been found. Darwin's theory, however, already contained a sort of response to this type of readings; at the beginning of Chapter IX, Darwin beckons for a more precise image to be formed of "what sort of intermediate forms must, on my theory, have formerly existed". Then, going on:

> I have found it difficult, when looking at any two species, to avoid picturing to myself, forms *directly* intermediate between them […] [But, this is] a wholly false view; we should always look for forms intermediate between each species and a common but unknown progenitor.[90]

For these same reasons, supposing an "ape-man" would be absurd from Darwin's point of view. This point was clearly remarked by Armand de Quatrefages, who based himself on Darwin's diagram (a fact that is noteworthy for its rarity): two related forms can have a common ancestor yet not be derived from each other, explaining why the ape-man theory cannot be Darwinian.[91]

Darwin's diagram does not start out with a single form but with ten, labelled A to L. So, although it does represent branching out from a root, the root itself is manifold, and the ancestors, from the outset, diverse. Darwinism's ambition is not to relate everything back to a single form by conjuring up the relationships between all existent forms. What must be remembered is the distinction drawn between the two meanings of "origin": Darwin's diagram describes origin as a mechanism of supervenience (*Entstehung*) but does not lead us to origin as a single source (*Urpsrung*). This aspect of the diagram, although but little highlighted, is nevertheless fruitful and is translated by the way it connects species together. Ludwig Büchner observed that the search for intermediary forms among presently existing forms is not at all a Darwinian one:

> It is an error, according to Darwin, because the forms currently in existence do not proceed one from the other, but each of them is the result and the last

term of a long series of developments. Thus, when one wishes to link two specific forms, it is *not* a direct intermediary that must be found for them, but some common ancestor that is unknown to us.[92]

This point is the source of much confusion: "For example, people say to you things like: *what! are you trying to convince us that a lion can come from an ass or an elephant from a tiger!!*"

And thus, farewell was bid to the derivation of the whale from the fly-swallowing bears. Understood in the above way, the diagram proposes a logic of always unknown progenitors—hence the ellipse-like dots at the bottom of Darwin's diagram. Those supposed progenitors weave together relations and give substance to the connections between given forms. From this point on, the absence of intermediaries or links between *existing* forms is no longer problematic: it is just that the common roots have disappeared, just as far distant into the past as the forms to be connected are dissimilar.

The examples of more or less fictitious genealogies that Darwin resorted to are a potent source of contention. It seems as though Darwin limits himself to indicating certain analogies between behaviours: a bear filtering water compared to a whale, or the titmouse breaking yew seeds with its beak compared to a nut-cracker. On such a basis, he suggests the idea of a *possible* transformation and of common ancestry. However, this mere possibility does not suffice, not even if the struggle for existence and natural selection are granted their full potentiality. Readers of the *Origin* could not help but conclude that the Darwinian argument was missing something.

This impression undoubtedly reaches its paroxysm in Chapter IV, where the principles of natural selection are presented. Darwin "must beg permission to give one or two imaginary illustrations", specifically his fleet and swift deer and wolves, pressed for food, becoming ever slimmer and swifter, season after season.[93] It seems that Darwin fully endorsed this "imaginary" aspect, one might even say proudly, given his frequent repetition of it.[94] In much the same way, the diagram in Chapter IV does not provide real genealogies, but rather an idealised frame for illustrating a mechanism. In both cases, Darwin's own conception ("my view") is presented to the reader as a hypothesis. This is not because there are not enough facts to prop it up, but because it illustrates a mechanism.

Indeed, what rare commentary the diagram did attract in the nineteenth century was invariably mocking in its attitude towards the absolutely hypothetical or "imaginary" character of the chart. For instance, Hyacinthe de Valroger writes in 1873: "M. Darwin has, in a chart, depicted these imaginary effects of natural election upon the descendants of a common parent".[95] Then, Valroger underlines the passage, "each interval … can, he says, represent a thousand generations, or better still, ten thousand; but it may represent a million, or even one hundred million". And he comments: "Just as many gratuitous, indemonstrable, indefensible assertions!" The diagram is criticised for providing no concrete application whatsoever, and for not employing data from taxonomy; further-more, much criticism was levelled at Darwin's constant use of fictitious examples. These fictions are all the more decisive for understanding what Darwin was attempting with the *Origin* that they constitute the very core of a general criticism claiming Darwin's masterpiece to be a mere accumulation of hypotheses rather than a work of science. As Darwin wrote to W.H. Harvey: "You object to all my illustrations: they are all necessarily conjectural, and may be all false; but they were the best I could give".[96] By this, Darwin shows that his interest lies more in deviation than in derivation. The *Ur*-ancestor is not his concern at all.

Whenever the diagram of Chapter IV does get mentioned in comments and reviews of the *Origin*, it is to compare it to linguists' diagrams.[97] The comparison always confers superiority upon the latter. The advantage with linguistic genealogies is that they involve proven relations between entities whose traces we do have. Darwin's diagram, on the other hand, is judged to be purely hypothetical. With a view on the works of linguists, Darwin's diagram was interpreted as an *Ursprung* rather than an *Entstehung*: Darwin's search for the *mechanism* was deemed secondary to the linking up of groups; the logic of the *common ancestor* was interpreted in terms of missing links. This led to pushing into the background the particular mechanism that Darwin highlights. Darwin's theory was being re-interpreted as merely a theory of development from a single germ, or as a theory of variation.

Notes

1 Darwin to Lyell, 2 September 1859, CCD 7 329.
2 *Origin* 1859, p. 459, *Var* 719 (#4).
3 Darwin to A. Gray, 11 May 1863, CCD 11 402–403.
4 *Origin* 1859, p. 199 and 201; *Var* 367 (#208) and 372 (#223).
5 *Origin* 1859, p. 236, *Var* 418 (#253). Darwin would eventually depersonalise this sentence in the fifth edition by exchanging "my theory" for "the theory" (#253:e).
6 *Origin* 1859, p. 206, *Var* 378–379 ((#267–271).
7 *Origin* 1859, p. 208, *Var* 381 (#13).
8 *Origin* 1859, p. 209, *Var* 382 (#25).
9 *Origin* 1859, p. 245; *Var* 425 (#8).
10 *Origin* 1859, p. 336, *Var* 547 (#190).
11 *Origin* 1859, p. 420, *Var* 656 (#81).
12 *Origin* 1859, p. 420, *Var* 656 (#79).
13 *Origin* 1859, pp. 457–458, *Var* 718 (#378) (emphasis added).
14 DNS, p. 280; *Origin* 1859, p. 133, *Var* 278 (#21).
15 Haeckel 1874, p. 92.
16 Darwin 1871, vol. 1, pp. 2–3.
17 *Ibid.*, pp. 3–4.
18 Lyell 1830. On Lyell, see Bartholomew 1973, Rudwick 1976, Corsi 1978.
19 Hodge 1977.
20 *Origin* 1859, p. 95; *Var* 185 (#124).
21 Mayr 1982, p. 478.
22 *Ibid.* See also Oldroyd 1984.
23 For a textbook presentation of the argument, see for instance, Chapter 3 ("One long argument") in Ruse 2008, pp. 54–74, and "The *Origin of species*", one of Ruse's own contributions to Ruse 2013, pp. 95–102.

24 Cf. Herschel 1831. Darwin to Bunbury, 9 February 1860, CCD 8 76; to G. Bentham, 22 May 1863, CCD 11 433. See Kavalovski 1974.
25 Herschel 1831, p. 149.
26 Cf. Hodge 1977, 1987 and 1989. All three of these major contributions reappear in Hodge 2008.
27 Cf. notably Ruse 1981 pp. 58–59. See also Ruse 1975.
28 Gray 1877, pp. 23–24.
29 Darwin to Henslow, 8 May 1860, CCD 8 195.
30 Himmelfarb 1959, p. 122. This passage is quoted (and rebuked) by Ruse 2006, p. 5.
31 Ruse 2008, p. 54.
32 Darwin to Baden Powell, 18 January 1860, CCD 8 41; Darwin to Lyell, 23 November and 10 December 1859, CCD 7 392 and 423.
33 Sedgwick to Darwin, 24 November 1859, CCD 7 396.
34 Mill 1872, vol. 1, p. 499.
35 Ruse 2008, p. 54.
36 *Origin* 1859, p. 159, *Var* 308 (#219).
37 Origin 1859, p. 352, *Var* 568 (#53).
38 *Origin* 1859, p. 482, *Var* 750 (#200).
39 I will return to this in Chapter 6.
40 Vorzimmer 1972, p. xvii.
41 Mayr 1982, p. 479.
42 On the actual plurality of arguments, see Mayr 1985; Sober and Orzack 2003; Waters 2009; Sober 2011 (Chapter 1).
43 Gruber 1981, p. 172.
44 *Origin* 1859, pp. 489–490, *Var* 758–759 (#268).
45 *Origin* 1859, p. 82; *Var* 166 (#22).
46 *Origin* 1859, p. 345; *Var* 561 (#267).
47 *Origin* 1859, p. 410: *Var* 645 (#201).
48 *Origin* 1859, p. 411, *Var* 646 (#7–8).
49 *Origin* 1859, p. 459; *Var* 719 (#5:e).
50 Darwin to J.D. Hooker, 27 March 1861, CCD 9 70.
51 Darwin to Baden Powell, 18 January 1860, CCD 8 39.
52 See Freeman 1977, p. 78; Johnson 2007.
53 Butler 1887, p. 204.
54 *Ibid.*, pp. 94–95.
55 Huxley 1893, vol. 2, p. 24.
56 Cf. Huxley (L) 1900, vol. 2, p. 192. For a recent defence of Romanes, see Forsdyke 2001, p. 38.
57 See below, Chapter 6 (on Romanes).
58 Becquemont 1992, p. 297 speaks of "Romanes or faithfulness to the letter".
59 *Origin* 1859, pp. 457–458. Huxley 1893, vol. 2, p. 232.
60 Mayr 1964, p. xxii.
61 A third example here could be Stephen Jay Gould's reaction to Darwin and the *Origin*. On this, see Gayon 2009b.
62 Mayr 1982, pp. 505–510; Mayr 1985.
63 See for instance, Mayr 1972, p. 64.
64 Darwin's five theories are, according to Mayr (1985): evolution as such, common descent, gradualism, multiplication of species, natural selection.
65 *Origin* 1859, p. 476.
66 Tort 1999.
67 Gervais 1859, p. 379.
68 *Origin* 1859, p. 343.
69 Darwin 1871, vol. 1, p. 1: "notes on the origin or *descent* of man"; and the title of the first part: vol. 1, p. 7: "Descent or origin of man".

70 Haeckel 1873, pp. 107–108.

71 Kellogg 1907, e.g. p. 17, section "theory of descent and the theory of natural selection distinguished".

72 Kellogg 1907, p. 30.

73 I follow the classical presentation given in the Summary of Chapter IV. *Origin* 1859 p. 127; *Var* 270–271 (#383–386). See also Darwin to Lyell, 30 March 1859, CCD 7 273.

74 Gayon 1998.

75 *Origin* 1859, p. 84; *1860*, p. 84; *Var* 168–169 (#40).

76 Young 1985.

77 Darwin 1868, Introduction, vol. 1, p. 6.

78 See Eiseley 1959a, 1959b and 1979.

79 *Origin* 1859, p. 62; *Var* 146 (#25).

80 The phrase serves as subtitle to Ruse 1981.

81 Darwin 1868, Introduction, vol. 1, p. 6. On this, see Hoquet 2013.

82 On unconscious selection, see Alter 2007.

83 *Origin* 1869, p. 72; *Var* 145 (#15.1:e).

84 Darwin to Asa Gray, 11 May 1863, CCD 11 403.

85 Jean Gayon has devoted many reflections and given several lectures on this topic. On the history of the diagram and tree-like representations, see Voss 2010. On Darwin's diagrams as a coral of life, see Bredekamp 2005.

86 Darwin to Lyell, 2 September 1859, CCD 7 329.

87 Cf. CDM, 164.

88 On the function of imaginary examples in the *Origin*, see, for instance, Griffiths 2016, Chapter 5.

89 *Origin* 1859, p. 184, *Var* 333 (#97). This passage is often mentioned by critics, e.g. *Blackwood's Edinburgh Magazine*, vol. 92, July 1862, p. 91.

90 *Origin* 1859, p. 280, *Var* 476 (#17).

91 Quatrefages 1867, p. 244: "It is not even necessary to read Darwin's book to be convinced that I am faithfully translating his ideas here; it is sufficient to cast one's eyes over the plate that depicts them graphically."

92 Büchner 1869, pp. 88–89.

93 *Origin* 1859, p. 90, *Var* 176–177 (#85–86).

94 *Origin* 1859, p. 90, 93, 94, 95; resp. *Var* 176 (#84), 182 (#112), 183 (#115), 185 (#124).

95 Valroger 1873, p. 93.

96 Darwin to W.H. Harvey, 20–24 September 1860, CCD 8 371–372.

97 The analogy with languages is proposed by Darwin himself: *Origin* 1859, pp. 422–423, *Var* 658–659 (#99–101). For recent accounts of the analogy between linguistic and organic evolution, see Van Whye 2005, Richards (RJ) 2013, Chapter 8.

Part II

Darwin-the-Variationist

In this part, I provide evidence that Darwin was never a pan-selectionist. In contrast to some of his followers, Darwin always supported the view that, besides natural selection, "other means of modification" had also been involved in the origin of species. This is especially clear in what Darwin states in a "most conspicuous position", at the end of the introduction to the first edition of the *Origin*.[1] Darwin always remained firm on this point. He never claimed that natural selection was the alpha and omega of evolution. This provided ample ground for various readings of the *Origin* to emerge, with readers like G. Romanes claiming to be the true heirs of Darwin against the "ultra" Darwinian views of Wallace or Weismann. The Darwin looking for "other means" may be called "Lamarckian" insofar as he was mobilised to counter selectionist accounts of evolution. Usually, "Lamarckian" evolution describes a theory where the strengthening or weakening of parts of the body is transmitted to the offspring and progressively raised to the rank of a specific feature. Lately, the *Origin of Species* has even been described as "The origin of species by means of use-inheritance".[2]

The efficiency of natural selection as the "means" for the origin *of species* was contested by some readers. Romanes, for one, understood that natural selection explains not the origin of *species* but merely the origin of *adaptations*. If this were truly the case, what would be an appropriate answer to the question of the origin of species: a question that Darwin perspicuously put to the fore, but left unanswered?

Variation may be an answer to this question. The first chapters of the *Origin* focus on the concept of variation. And while the struggle for existence and natural selection are indeed essential components of the Darwinian edifice, they are nevertheless duly and tightly contained within the variation committed grip of Chapters I, II, and V. The Darwinian system can be defined as the non-random survival of random hereditary variations. Variation is the harbour from which Darwin sets sail. This is only to be expected from a work that treats origin as an *Entstehung*, a mechanism of origination. If Darwin's theory is the theory of "descent with modification", then just how modification supervenes in lineages— by means of variations produced during reproduction—is an essential axis of examination. Darwin's theory provides the basis for an important programme of research: on the origin of variations. This point is especially clear in the conclusion of the *Origin* (Chapter XIV) when Darwin explains what should be sought for by future generations: not an experimental programme to test whether natural selection really does operate in the wild, but a programme fixed "on the laws of variation". Variation is Darwin's prime target when referring to "a grand and almost untrodden field of inquiry".[3]

Notes

1 See *Origin* 1872, chapter XV, p. 421.
2 Waller 2002.
3 *Origin* 1859, p. 486.

4   Darwin-the-Epicurean
Chance and laws of variation

Darwin's theory of variation has been the object of various theoretical concerns, especially when it comes to "indefinite" variation and the role of chance.[1] As a result, the *Origin* has often been read as an Epicurean treatise: both because of the place it grants to chance (the randomness of variation) and for its vision of nature as a field subjected to some blind force (natural selection). The Darwinian world is seen as Epicurean insofar as it excludes any kind of teleology and erases all notion of intelligence from the foundations of the world. This is why the *Origin* became the brunt of so many objections which had classically been levelled at Epicureanism by Providentialist philosophies: how could the blind and aimless accumulation of tiny random variations stacked on top of each other possibly create order, beauty, intelligence, in a word, a *kosmos*? Those who consider the *Origin* to be just another Epicurean apologetics of "chance" can attempt either to reconstruct it as a materialist system or refute it by discovering principles of order and laws of nature.

This view of Darwin's *Origin* as a new form of Epicureanism was especially popular in the popular press. As such, the *Origin* was understood to be in contradiction with William Paley's argument from Design. However, Darwin's stance on chance is more complex than it appears. First, as Ellegård aptly put it:

> By declaring that the sorting out of the viable from the non-viable was made on the basis of innumerable minute variations as between parents and offspring, and not on the basis of organisms arising complete out of the chance collocation of elemental atoms, Darwin turned the Epicurean hypothesis from a nearly absurd into an eminently plausible one.[2]

Besides, Darwin gives a very strange form of Epicurianism, if any. According to Alexander Grant, Darwin's *Origin* "is the theory of Epicurus", but "with the atheism removed".[3] How then is it still an Epicureanism?

In contrast to "chance variation", the vast majority of Darwin's contemporaries and readers insisted upon the lawfulness of variation. Enemies of chance would often quote Darwin's own admission that the expression "mere chance" is nothing more than a make-do, a vague manner of speaking of our ignorance as regards true causes.[4] They concluded that Epicureanism is impossible and turned

Darwin into the discoverer of a new field of phenomena whose laws awaited only to be uncovered.[5] Thus, the partisans of natural laws indicated that chance, of the sort advanced by Darwin, did not exclude law or determination; chance is not a reality but merely a word that signposts (potentially provisional) ignorance, perhaps even the very sign of a contemporary incapacity to adequately describe the laws at work behind the origin of such and such an event.

Darwin's emphasis on a great abundance of non-adaptative and unpredictable variations allowed his opponents to consider that the fitting variation would be beyond the reach of chance: "Darwinism" faced the same standard objections that had always been levelled at Epicurean philosophy. In 1867, Fleeming Jenkin asked of Darwin how one and the same variation could arise simultaneously in numerous individuals. In 1871, St. George J. Mivart brought up "the incompetency of 'natural selection' to account for the incipient stages of useful structures" and structural co-adaptations between two interdependent individuals: variations would have to simultaneously affect the milk producing mammary in the mother as well as the suction and digestion organs in the offspring.[6] But, as early as 1860, Richard Owen was asking "if certain bounds to the variability of specific characters be a law in nature", or if "unlimited variability by 'natural selection' be a law".[7] That variability exists is undeniably established by the differences in individuals; indeed, it is the very condition of individuality, but is this individual variability a sufficient condition for the transformation of species into genera, classes, and so on?[8] Owen did not ignore natural selection, but he made its action depend on the production of individual variation. Variability, and specifically its limits or its infiniteness, appear to be primordial, forming an indispensable precursor to the agency of natural selection.

Turning Darwin into an Epicurean (Eimer and Berg)

We will now look at two readings of the *Origin* that make an Epicurean of Darwin: those of Theodor Eimer and Leo S. Berg, both of whom bemoaned the exaggerated place the Darwinian system dedicated to chance variation.

Theodor Eimer constantly endeavoured to cloak himself in Darwin's authority, to the point that he even called his own book *Entstehung der Arten*, i.e. "Origin of Species".[9] He openly faults Darwin for not achieving his avowed goal while challenging him on that very same ground. Eimer bent the *Origin* to his own hypothesis, opposing other interpretations of Darwin that stand accused of overly emphasising the principle of utility. He considers that the *Origin* exposes a theory of chance and therefore tends to separate Darwin's words from any consideration of utility. Thus, his reading aims to complete Darwin by seeking laws, although he contests the rule of the supposed "*Darwin'sche Nützlichkeitsprincip* (Darwinian Principle of Utility)". In Eimer's view, Darwin never defended such a principle and, in fact, only further distanced himself from it with time. In this regard, Eimer notes that Darwin himself regretted certain turns of phrase which tended to pull all that exists under the dominion of utilitarian criteria.

The central aim of Theodor Eimer's own *Origin of Species* was to establish whether or not variation is law-like. From his perspective, Darwin was at fault for giving himself completely over to a theory involving variations "without rules": instead, what was needed was to show the "lawfulness of variation, contrary to the rule of pure chance as asserted by the Darwinian theory". Nevertheless, what Eimer understands to be the spirit of Darwinism is not entirely incorrect: its philosophical foundations consist in the search for natural lawfulness and can, therefore, be tested. For Eimer, the very project of Darwinism is to extend lawfulness so as to include those very points Darwin had (provisionally?) left to chance. So, it is not a case of refuting "Darwinism" but rather, by adopting a stance of genuine commitment to it, extending the search for lawfulness to an area that Darwin had neglected and left in darkness: the origin of variation, that is to say, its causes and laws. Eimer proclaims his stance to be legitimated by the very text of the *Origin*: his approach can base itself on the idea that a philosophical worldview called "Darwinism" has its own logical coherence, which needs to be established beyond the actual content of Darwin's *Origin*. Chapter V of the *Origin* contains just such an invitation to extend Darwin's explicit lesson by further exploring the question of the laws of variation.

All things considered, Eimer is extremely clear: his objective is not to lay waste to Darwinism (the system he thinks best captures the meaning of Darwin's texts) but actually to reinforce it: "If we show that the origin of variations is lawful, then this justification will also apply for the origin of species".[10] And this is not a simple exit clause. It is drawn from the very lessons derived from "Darwinism" and the theory of evolution in general (*die Entwicklungslehre*): that there is no essential difference between species and sub-species or varieties. According to Eimer, to the extent that, in principle, the differences between species are not distinct in nature from the differences between varieties, or, more so, since species result from individual variations, so establishing the lawfulness of individual variations reinforces the Darwinian idea of there being a lawfulness to the origin of species. Eimer criticises the *Origin* for leaving two questions in suspense, questions which open up a "huge gap" at its core: the lawfulness (*Gesetzmässigkeit*) and causes (*Ursachen*) of variations. The most fervent followers of Darwin advanced an active power: "a selection (*Auslese*) which relies on the useful". But they cast no light on the production of variations in the characters of individuals, in equal parts the very material of selection and that which selection does not explain. Rather, it explains only their rise and domination ("*die Steigerung und das Herrschend*"), not their origin.

The Epicureanism objection crops up constantly, across various readings of the *Origin*. This is a clear reflection of the obsession its readers have had with laws. Theodor Eimer claims that in submitting variation to laws he is faithfully continuing Darwin's own work, although he is quite singular in this perspective. Most often, readers identify the *Origin* with an apology of chance variation. Once this is affirmed, the discovery of laws, far from continuing Darwin's work, completely upturns his thought. Without chance, Darwin is torn away from the foundations of his thought. This is the argument defended by Leo S. Berg in a book aptly called "*Nomogenesis*", a term whose meaning is revealed by the sub-title: "*evolution determined by law*". Berg aims at bringing the laws of evolution to the fore and, through this, at cancelling out "Darwinism" (i.e. a world view that equates nature and chance). Darwin finds himself compared to Diderot in the latter's *Pensées philosophiques* and made the object of the classical anti-Epicurean objection of the infinite dice throws which could never produce the *Iliad*.[11] For Berg, the word "law" in the natural sciences means the exclusion of chance and he duly proceeds to pronounce several such laws. He continues the work of several Russian critics, for whom

> any newly defined relation, new law or rule in the variation of organisms, in the course of heredity or the phenomena of hybridisation and reproduction, which we may discover, *cancels* the theory of Darwin. For the Darwinian process requires as a necessary condition a complete indefiniteness in all these processes, a veritable chaos, from which order must emerge of itself, under the effect of one determined principle, that of utility, or preservation from destruction.[12]

In his criticism of Darwin, Berg fully understands the logic behind his philosophy:

> Darwin assumes that the variability of organisms is so great that chance, in adapting characters (in a manner analogous to that in which the opener of the lock fitted his keys), will always select an accidental variation which may prove useful. Selection operates on accidentally useful variations.[13]

The difference between Berg and Darwin is illustrated by their respective inter-pretations of divergence and convergence. For the Darwinians, differences in the structure of organisms are "the result of a divergence, i.e. a deviation of charac-ters in different directions, due to variation, to accidental utility, to the struggle for life, and to the survival of the fittest"; as for the similarities, these are inter-preted as a fundamental element, the consequence of characters passed down from the common ancestor. Conversely, in an approach very close to the one Bergson adopts in his *Creative Evolution* (1907), Berg notes that common char-acters do exist between groups so different that these cannot possibly be the result of inheritance from a common ancestor. For instance, the similarity between whales and fish, or dolphins and ichthyosauruses. For Berg, "conver-gence of characters" is "really the rule", and it is "inconsistent with the principle of chance: it is a result of development in a determined direction".[14]

This is why Berg turns the Darwinian view upside down and maintains that the differences do not stem from divergence in characters but are the effect of inheritance from common ancestors; it is the differences (and not the similar-ities) which are aboriginal. Rebutting all distinction between homologies and analogies, he adds that, very often, similarities are not the result of inheritance from a common ancestor but a consequence of the convergence of characters. If it is true that, in evolution, we find both aboriginal differences and differences acquired through divergence, as well as aboriginal resemblances and resemblances which result from convergence, then "the general trend of evolution, its primary traits, are due to aboriginal differences and convergent similarities, and this is inconsistent with the theory of natural selection".[15] On this basis, the fact that Berg makes reference to D'Arcy Thompson's work is understandable, as too is the fact that the latter wrote the preface to the translation of Eimer's book; D'Arcy's work shows that resemblances are the product of identical morphological constraints and not the mark of a common ancestry.[16]

From chance to prediction: the unknown causes of variation

The baseline of these interpretations of Darwin as an *Epicurus redivivus*, is Darwin's constant effort to widen the range of variation. Asking whether "future varieties of wheat and other grain [will] produce heavier crops than our present varieties", Darwin gives this answer in his *Variation*: "These questions cannot be positively answered; but it is certain that we ought to be cautious in answering by a negative". On the one hand, "in some lines of variation the limit has probably been reached", but on the other hand,

> seeing the great improvement within recent times in our cattle and sheep, and especially in our pigs; seeing the wonderful increase in weight in our poultry of all kinds during the last few years; he would be a bold man who would assert that perfection has been reached.[17]

Further still, Darwin refutes any fear that variability might exhaust itself. He quotes the French agronomist Augustin Sageret (1763–1851): "that the most important principle is 'that the more plants have departed from their original type, the more they tend to depart from it'. There is apparently much truth in this remark". It may be that the path followed, from varieties of wild fruits to cultivated fruits, was very long, but certain facts can, as Sageret put it: "console us and give us the hope of speeding up the process".[18]

This reference to Sageret is of interest in itself. Sageret, in his *Pomologie physiologique* (1830), had launched a genuine appeal for an understanding of the laws of variation to be reached in order that man then be able to reproduce them *ad libitum*. Endeavouring to lay down the foundations or principles of such a science, he wrote that success in agriculture relative to "the improvement of species as regards flavour and fruit quality, increase, acceleration and the certainty of the produce, acclimatisation of foreign species, creation of varieties and, even more importantly, as regards the direction this creation should follow", was all for the most part "due to chance, that is to say, a fortuitous conspiracy of circumstance, perhaps, but in any case not led and not predicted". He lamented the fact that "there is no doctrinal body to inform us, not even to point us in the right direction and guide us in this research".[19] Sageret wanted to have pomology established as a science and thereby eliminate chance in two ways: our knowledge of causes would lift the obligation to speak of "chance" in place of the genuine causes of variation. Additionally, we would no longer be powerless and blind with respect to the production of these variations and would, in fact, be able to actively reproduce them. Causal explanation, but above all predictability of variation and the capacity to reproduce it at will: this is what marked the radical difference with the situation of horticulturists who, while "advanced to a certain degree in practical Pomiculture", were nonetheless "absolutely in the infancy of the art as regards theory and the improvement of the species". So, what was to be expected from the birth of pomology? Above all, the means to ensuring "the perfecting of fructification [carpophore]". But, ultimately, Sageret went beyond this and formulated what would later be Darwin's own question: the origin of species through discovery of the unknown causes of variation. Sageret declared himself incapable of founding this doctrine of agriculture and therefore limited himself to proposing "a collection of facts, experiments, observations, comparisons, conjectures and reflections whose connection and organisation remain to be set out". Agriculture was just waiting for the philosopher capable of organising these patiently collected and compared facts. Darwin, who quotes Sageret on the subject of variation, hoped that it would now be possible to accelerate the transformation of domestic species. The time was therefore ripe for a transition from "chance" to genuine science, from passive cultivation (awaiting the emergence of favourable traits) to active and predictive cultivation (producing the variations needed). Darwin may well have believed himself to be just the man for the job.

Describing random variations (spontaneous, undetermined, limitless, non-adapted, unpredictable) is an invitation to search for their causes. The historical inquiry into variability cries out to be completed by an etiological dimension. Indeed, whatever about the magnitude of variations or their apparent undeterminedness, this does not anticipate on the regularity with which they are produced. Darwin himself indicated that we speak of *random* chance primarily due to our ignorance regarding causes and laws. Chance, then, is just the name that covers all unknown causes of variation. So, it turns out that the characteristics which qualify Darwinian variation as random do not exclude it from being governed by a certain causality. Undeterminedness may only be a veil for ignorance: it accompanies the progress of science and positively reinforces the space implicitly defined by the (provisional?) limits of our understanding.

Given that Darwin's intention was to discover the laws of variation, what would become of the "freedom" of variation should he succeed? How should the spontaneity of some organ's variation be understood if that variation were subject to various factors or laws—necessary correlation with the variation of other organs; strong tendency to vary proportionately to the degree of development; or variations undergone in recent time…? Darwin was conscious that forces hid behind variation and its possible laws that may guide it, such as, notably, tendencies to reversion. These tendencies shape variation, countering the effects of selection.[20]

Variation can, and often does, consist in the reappearance of very old characteristics, rather than in the appearance new ones. Thus, to explain this, it must be admitted that "there is a *tendency* in the young of each successive generation to produce the long-lost character, and that this tendency, from unknown causes, sometimes prevails".[21]

As is well known, alongside the principle of divergence laid out in Chapter IV, Darwin also posits a "tendency to depart". And it seems like he again supposes the existence of a tendency to vary when, for example, he writes: "Such considerations as these incline me to lay less weight on the direct action of the surrounding conditions, than on *a tendency to vary*, due to causes of which we are quite ignorant".[22] But can we go from this *tendency to vary* to the tendency of variation to follow certain lines? Is the leap so great? If such were the case, could there be directions in variation independent of the divergence effect produced by selection?

The fifth edition of the *Origin* attempts to shed some light on this opposition between directed variation (centripetal, reversing) and free or undetermined variation (centrifugal, diffuse), indicating the possibility of variations affecting a whole population. In a rewritten passage of Chapter V, Darwin specifies that the direct action of a changed condition of life will lead to either indefinite or definite results. When the direct action of the conditions produces indefinite results, "the organisation seems to become plastic, and we have much fluctuating variability"; but, when the results are definite, "all, or nearly all the individuals become modified in the same way".[23] The "*tendency to vary*" is essential to the Darwinian perspective, because Darwin sees in it the real explanation for specific differences, the *vera causa* affirmation of the parental community, as opposed to the theory of independent creations.[24] Throughout Chapter V, Darwin affirms the necessity of moving beyond randomness by seeking the causes and laws of variation. However, since he also affirms that variation is most clearly manifest in domestication, breeding lots provided him with a particularly pertinent setting for the study of this phenomenon. Based on this, Darwin was able to attribute diverse causes to variation: a direct or indirect effect of the conditions of life, habit, use and disuse, correlation of growth, compensation or balance.[25]

Scope and breadth of selection

The quest for the causes and laws of variation was an obsession with Darwin. He returns to it endlessly, this dark cloud wherein the mobilising strings of the species-originating mechanism he discovered are pulled. He does manage to discern certain partial results but always ends up back at the same point: our ignorance of a complex phenomenon and the veil of "chance" we dress this ignorance up in.

In the view of many historians and biologists, Darwin opened a Pandora's box he would never again be able to close by wading into the laws of variation, unless, of course, he should find a satisfactory theory of inheritance. This is why the years from 1859 to 1882 are often presented as a long period of contestation and controversy which only truly came to an end with the emergence of a "good" theory of inheritance in the 1900s, the crown of which was the neo-Darwinian synthesis of the 1930s and 40s.

On the other hand, we can take Darwin at face value and read the successive alterations he brought to his theory not as contortions or impoverishments, but as genuine enrichments. Abandoning the retrospective point of view the biological synthesis of the 1940s injected into the history of Darwinism is one way to take cognisance of Darwin's method and to observe the sheer delight he took in propping up his system with ever more recent facts, in particular those linked to domestic animals and plants. The history of the *Origin*, as revealed in Peckham's study of the *editio variorum*, tells us that, far from fixating on natural selection's omnipotence, Darwin confronted the most robust objections head on, keeping his mind ever open to new facts of which he was made aware.

Granting central position to variability and the laws of variation results in agriculture and teratology being placed at the very heart of Darwinian reflection. All reasoning on variations makes unending reference to the context of either cultivation (vegetable or animal) or teratology, which attempt to establish themselves as predictive sciences. In this context of the investigation of natural laws, what meaning can "random" variation possibly have? We have seen that Darwin did not necessarily support the idea of "random" variation since, on the contrary, his objective was to determine its laws. On the characteristics of variation, he gave consideration to those objections which tended to limit the quantity or determine the directions of variation; on causes, given that the term "chance" labels only our ignorance, Darwin actively sought out the laws of variation and the causes of variability.

This question about the nature of variation's actions impacted the way the *Origin* was initially read and communicated. Early readers, disciples, and critics all placed variability at the heart of the Darwinian system and prioritised research into the laws of variation. Further still, they related the debate over the value of natural selection to the debate over variation, under the double aspect of its magnitude and its lawfulness. This field fed into an important series of objections against Darwin's theory which suggest pushing the question of variability ahead of the question of natural selection. Variability, deemed to be either equivalent or more important, was in any case systematically granted precedence. In 1874, Albert Wigand, for instance, formulated this in a way which plainly reveals the shift that was taking place, when he mused about "what selection (we could just as well say: variability) can effect".[26] Selection sees its agency curtailed by far-reaching variability. This argument leans on Darwin's own words when he equated "quantity of variation" with "limit of selection".[27]

That causes and laws of variation exist, and that we are largely ignorant of their nature, does this necessarily constitute an obstacle to the theory of the origin of species? Certainly, Darwin seems to have been of this opinion when he wrote to Hooker in 1862: "my present work is leading me to believe rather more in the direct action of physical conditions. I presume I regret it, because it lessens the glory of Natural selection, and is so confoundedly doubtful".[28]

Variation's non-random character would disarm the explanatory virtue of natural selection in two different ways. On the one hand, if variation is neither undetermined nor limitless, if, on the contrary, it imposes strict boundaries, then this reduces the *scope* of natural selection; it finds itself confined within certain such boundaries. On the other hand, if variation is not random, but follows determined directions, then this reduces natural selection's *breadth* and seems to channel it in certain directions only.

Remarkably, the general interrogation does not relate to the existence of natural selection but rather to the constraints which weigh upon it *qua* agent—constraints antecedent to its action and with which it must create. Thus, the two questions of the magnitude and direction of variations do not occupy exactly the same level (Table 4.1). The first question bids us think that natural selection cannot account for everything. In this, it reduces selection's field of adequacy. The second objection, though, indicates that natural selection must fall into certain directions which it does not choose. Here, it questions selection's responsibility for the transformations that are actually observed.

Finally, what relationship can be established between Darwin's *Origin* and lawfulness (*Gesetzmässigkeit*)? An American defender of Darwin, Chauncey Wright, got himself quite worked up over the recurrence of this question, the flip side of the Epicurean dispute.[29] We are in the nineteenth century, he insisted on reminding people; nobody contests the importance of causal determinations any more! If Darwin is accused of Epicureanism, it is because his readers are searching for *laws*. However, either they must consider Darwin's system to be the dominion of chance, in which case the discovery of laws, far from extending Darwin's work, will instead lead to the total subversion of Darwinian thought, now torn asunder from its foundations; or else, on the contrary, they will have to integrate Darwin into the project of searching for laws (such as natural selection), in which case he cannot be an Epicurean. Thus, by proposing other laws for explaining the theory of evolution, Darwin's successors claim to do nothing other than extend or complete a scientific edifice which, far from being set in stone, calls for the discovery of other lawful relationships and connections. If scientists must search for laws, then natural selection does not suffice and Darwin will have to be overtaken on the very path he himself marked out. Darwin did advance a law (natural selection), but he also left important pages of

Table 4.1 Effect of the characters of variation on the efficiency of natural selection

Character of variation	Effect on natural selection	Aspect of natural selection questioned	Darwinian response
Limited	Bounded	Its scope, its magnitude (Adequacy)	Accumulation of a multiplicity of infinitesimal and undetermined variations
Guided	Channelled	Its breadth (Responsibility)	Exterior circumstances form only a frame

natural science blank, leaving room for "chance" i.e. ignorance, especially in all that involves the origin of the *fittest* or the origin of variations. This is why his work must be completed by a search for the laws of variation. In accordance with this second attitude, every scholar proposes new laws as a contribution to the edifice of the theory of evolution. But what value do they have? Bouncing back and forth between the selectionists to one side and every other movement to the other, we see something like the opposition between the Cartesians and the Newtonians being played out anew. The former were determined to hold firmly onto impulse by contact only, considering the latter's "attraction" to be a purely occult quality. Likewise, the new laws put forward in order to complete selection were often deemed to be unnecessary or unsupported by the facts. So, when the "Neo-Lamarckian" Edward Drinker Cope referred to some new (and quite mysterious) "law of polar or centrifugal growth", the "Ultra-Darwinian" Alfred Russel Wallace referred instead to mere variations "thus preserved and increased by natural selection".[30]

But, from the point of view of those who were advancing new laws, natural selection suffered from being only a *local* law, an ad hoc principle, explaining order only in the *organic* world. And yet the inorganic world, where natural selection has no dominion, does not lack order. It would, perhaps, be therefore more fitting to explain all organic phenomena by means of the inorganic (principle of reduction) or, at the very least, to see what portion of organic phenomena can be explained using the laws of the inorganic. In order to avoid accusations of Epicureanism, Darwin's project could be redefined as the search for natural laws and, through this new extension, could be reoriented. In particular, the search for the laws of variation implied no longer evoking *chance* but rather *determined* variation. As Huxley notes: "Darwin has left the causes of variation and the question whether it is limited or directed by external conditions perfectly open".[31]

Notes

1 See Gigerenzer *et al.* 1989, pp. 132–141; Winther 2000; Beatty 2006; Merlin 2013.
2 Ellegård 1958, p. 116.
3 Grant 1871, p. 281. See also, in Ellegård 1958, p. 116, references to *John Bull*, 24 December 1859, p. 827; *Month*, 11 (1869), p. 289.
4 *Origin* 1859, p. 131; *Var* 275 (#4–5). This text would later be shored up by a passage from the final chapter to Darwin 1868, vol. 2, p. 420.
5 See Jaeger 1860, p. 99: Darwin "proves that a lot of what we were accustomed to dismissing under the heading 'Chance' is as law-like as the currents in the atmosphere and the ocean". See also Argyll 1867, p. 255 or Berg 1926, p. 23.
6 Jenkin 1867; Mivart 1871, Ch. II.
7 Owen 1860a, p. 520.
8 *Ibid.*, p. 521: "This true and proved law of variability is, in fact, the essential condition of individuality itself."
9 Eimer 1888. See Bowler 1988, especially pp. 151ff.
10 Eimer 1888, vol. 1, p. 1.
11 Berg 1926, p. 36.
12 N.N. Strakhov (1889), quoted in Berg 1926, p. 110.

13 Berg 1926, p. 37.
14 *Ibid.*, p. 157.
15 *Ibid.*, p. 157.
16 Thompson 1917.
17 Darwin 1868, vol. 2, p. 242.
18 Sageret 1830, p. 106, quoted Darwin 1868, vol. 2, 241.
19 *Ibid.*, p. 2.
20 *Origin* 1859 152–153, *Var* 301–302 (#174).
21 *Origin* 1859 p. 166, *Var* 316 (#277).
22 *Origin* 1872, p. 107, *Var* 279, #32:f. (emphasis added).
23 *Origin* 1869, p. 166, *Var* 276 (#12.4:e and 12.5:e).
24 *Origin* 1859, p. 159, *Var* 308–309 (#219).
25 Chapters XXIV to XXVI in Darwin 1868 are devoted to the issue of the laws of variation. Vorzimmer (1972) analysed Darwin's persistent attempts to investigate the laws of variation.
26 Wigand 1874, vol. 1, p. 56.
27 Darwin 1868, vol. 2, p. 242.
28 Darwin to Hooker, 24 November 1862, CCD 10 556.
29 Wright 1871.
30 Wallace 1889, p. 422.
31 Huxley to Hooker, 23 March 1888, in Huxley (L) 1900, vol. 2, p. 205.

5 Darwin-the-Teleologist

Are all variations useful?

In contrast with Darwin-the-Epicurean, the *Origin* may also be reconstructed as so replete with utilitarian considerations that it is turned into a teleological system. Here, Darwin's philosophy is first seen as embodying in the natural sciences what John Stuart Mill's philosophy embodies in practical philosophy and philosophy of knowledge: utilitarianism, a simple by-product of Victorian society. In the Epicurean view of Darwin, chance and utility work together to exclude final causes and the beauty of the world. If the *Origin* is an Epicurean manifesto, then it refutes the existence of a rational, ordering cause of phenomena; if it is a utilitarian missal, then it recognises the existence only of what is useful and denies the possibility of all free (as opposed to "dependent") beauty. In both cases, Darwin's book equally embodies the negation of Providence and that of natural harmony. Darwin himself even went so far as to affirm that, had organisms been created only for the pleasure procured from their beauty, then such an objection would be "fatal to [his] theory".[1] But, strikingly enough, there are *other* ways of viewing the *Origin*'s combination of chance and utility, each of them posing specific problems to the coherency of what is understood to be the true Darwinian doctrine (see Table 5.1).

Table 5.1 Chance and utility in the *Origin*

Does the Origin support random variation?	Does the Origin support utility (effect of natural selection)?	Implications for the coherency of the Origin	
Yes	Yes	Random variation and utility work at two consecutive levels: first random variations are brought about; then, what is useful is naturally selected	
Yes	No	The *Origin* as Epicurean treatise	The origin of variations question is brought to the fore (whether random or according to laws). Natural selection is pushed into the background
No	Yes	The *Origin* as a teleological treatise	

The first line of Table 5.1 gives the typical reading of the *Origin*: Darwin affirms the principle of random variation and the principle of utility, guaranteed by the action of natural selection. However, these two dimensions may also be seen as opposing each other, insofar as one affirms the randomness of variation while the other refutes it by submitting everything to the reasoning of utility. How can coherency be re-established? Affirming chance variation and denying utility is the core of Epicurean readings of Darwin, such as Theodor Eimer's. But one may also devise a reading of Darwin as a teleologist, putting the emphasis on utility, and pushing chance variations to the background. Such paradoxical readings of Darwin as a staunch supporter of teleology are the object of this chapter.

A two-step mechanism

The two principles of random variation and utility can be reconciled as long as *finality* is distinguished from *teleology*: the former observes a function *ex post*, the latter assigns one *ex ante*. Where teleology gives itself a determined end in the form of an antecedent programme, finality limits itself to a diagnosis of consequent utility, a function undertaken by some structure which was not conceived of in view of that effect.

Natural selection is there to give shape and form to free and undetermined variability. In an addition to the sixth edition of the *Origin*, Darwin makes this précis: "But structures thus indirectly gained, although at first of no advantage to a species, may subsequently have been taken advantage of by its modified descendants, under new conditions of life and newly acquired habits".[2] This point is expressed wonderfully in the beautiful ending to Chapter XXI of *Variation*: natural selection is like

> an architect ... compelled to build an edifice with uncut stones, fallen from a precipice. The shape of each fragment may be called accidental; yet the shape of each has been determined by the force of gravity, the nature of the rock, and the slope of the precipice,—events and circumstances, all of which depend on natural laws; but there is no relation between these laws and the purpose for which each fragment is used by the builder.

Hence, Darwin concludes,

> though variability is indispensably necessary, yet, when we look at some highly complex and excellently adapted organism, variability sinks to a quite subordinate position in importance in comparison with selection, in the same manner as the shape of each fragment used by our supposed architect is unimportant in comparison with his skill.[3]

Darwin clearly differentiates between the two levels: the elaboration of raw material level (variability and its blind laws) and the more fundamental level of the edifice's construction (natural selection's use of these raw fragments). This is why, "Over all these causes of Change I am convinced that the accumulative action of Selection, whether applied methodically and more quickly, or unconsciously and more slowly, but more efficiently, is by far the predominant Power".[4]

In this same sense, the botanist Carl Nägeli affirms that Darwinian variation is undetermined (*unbestimmt*) but that this in itself does not justify labelling its world view as Epicurean; all the more so that Darwin's theory accounts for utility on a second level. Nägeli reads the *Origin* as a theory of "usefulness" (*Nützlichkeittheorie*) by distinguishing it from teleology (*Zweckmässigkeit*).[5] Since variations are produced purely randomly, and *not* for their utility, since, therefore, it is only *after the fact* that chance variations prove to be useful, it is thus quite impossible to identify any form of teleology.

A similar interpretation is given by both Thomas H. Huxley and August Weismann. Both were opposed to the hypothesis of creations, which takes everything to be entirely teleologically led (*zweckmässig*). But they also opposed the transmutational hypothesis, which allows for structures that have no function.

For his example, Huxley takes the teeth of the foetal *Balœna*: "Darwinically ... every detail observed in an animal's structure is of use to it, or has been of use to its ancestors"; in contrast, "teleologically ... every detail of an animal's structure has been created for its benefit". In other words, foetal teeth have meaning "Darwinically" speaking, but not "teleologically" speaking.[6]

Weismann's approach is to insist heavily on the uselessness of certain parts. He evokes "the rudimentary scapular belt of slow-worms and other lizards" and "the rudimentary pelvic belt of the cetaceans", stating that: "Nobody could go so far as to claim that these parts of their organisms are useful to them in any way whatsoever".[7] The *Origin* can easily account for uselessness, whereas it remains a mystery within the fully finalist vision of the creations hypothesis. Naturalists who refer to "the Creator's plan" imagine the independent and simultaneous production (*Erschaffung*) of all species; hence, Weismann says, they have to "acknowledge that there exist relations within the organic world and, alongside this, renounce on providing any kind of explanation for these relations". The supporters of transmutation, on the contrary, start with only a few of the simplest forms possible (*wenige allereinfachste Organismenarten*); they understand relations between forms as the result of resemblance through descent. And to the naturalists who refer to a hypothetical vertebrate "body plan (*Bauplan*)",

Table 5.2 Creation vs transmutation (according to Weismann 1868)

Creation	Transmutation
Immutability	Mutability
"The Creator's plan"	Relations, natural system of classification
"Species" is an absolute notion	"Species" has relative value only
Bauplan	Rudimentary organs

Weismann retorts by asking why the serpent, for example, lacks these rudimentary organs, given that it is so close to the lizards. If the answer given is that these characteristics do not belong to the specific body plan (*speciellen Bauplan*) of serpents, which have no limbs, then another fact must also be confronted: why does the group of giant snakes (*Riesenschlange*, including pythons and boas) possess "the rudiments of hind extremities"?

For Huxley, just as for Weismann, functionless rudimentary organs make sense within the transmutational theory: they are the leftovers of some former utility. Nevertheless, the existence of functionless structures notwithstanding, to characterise the *Origin* as a *Nützlichkeittheorie* is entirely justified when viewed through the lens of natural selection. Thus, chance and utility can be reconciled within the economy of Darwin's views, as long as one dissociates the production of variation and natural selection levels. The utility of characteristics (their adaptive nature) must be dissociated from the origin of variations. Even if characteristics arise independently of natural selection (produced by modifications in the environment, for example, or by some internal law of the organism, following a disturbance affecting the sexual organs or some such similar event), they can still end up finding some subsequent and indirect utility.

This classical reading would later be taken up by the disciples of the Modern Synthesis theory. Ernst Mayr, for instance, distinguishes the *blind chance, single-step* Epicurean model (which he deems to be non-Darwinian) from the *two-step process* which explains adaptations by means of the *fortuitous or calamitous* sorting of phenotypes.[8] Elliott Sober has even suggested that only creationists advance the caricature associating evolution and *randomness*. In other words, they have a tendency to read the *Origin* as though it were an Epicurean theory, something that it is not.[9] The random or undetermined nature of variation is no obstacle to the action of natural selection, if by the latter we mean a certain ordering that assures the production of what is useful. As Weismann very expressively put it in 1868: "It is therefore the action of the struggle for existence upon the variability of species, that is to say, natural selection, which occasions the origin of new races".[10]

Taking randomness as its starting point, the *Origin*'s tour-de-force is in managing to recoup all the results of classical teleology: in short, it manages to keep a hold on finality when it is present while still accounting for all that is devoid of manifest finality. Darwin cracks the old teleology chestnut and reduces it to a semblance: "*mechanistic purposiveness*" or simply the "*as-if-it-were-designed-by-God*" appearance of the animal kingdom.[11] Ruse summarises this point by saying that, "although the problem of final cause was certainly shifted and changed by the *Origin*, it was not obviously expelled or anathematized".[12]

This is all well recognised and would not need to be recalled in such detail were it not for certain critics of Darwin suggesting that this reading be reconsidered through an attempt to dissociate the two principles of chance and utility (whose conflicting agency is represented in the lower lines of Table 5.1, above). Indeed, several authors conclude that Darwin uses chance to explain everything and therefore neglects to consider any principle of utility or law; they make an

Epicurean of Darwin. Others, in contrast, associate the *Origin* with a plea for utility in nature and take quite improbable paths to arrive at denying chance any role within Darwin's conceptions; they make a *teleologist* of Darwin. The former disregard and the latter underline that variation itself is not produced by chance but that, as Darwin had indicated in Chapter V, it is subject to laws. Both of these readings favour just one of the *Origin*'s two principles, each to the detriment of the other. It could be said that, after a fashion, these interpretations fail to grasp the particular combination of chance and utility proposed within the Darwinian schema (reconstituted above). Thus, theirs would be more accurately termed as misinterpretations. However, they still hold interest as interpretative prisms: they distract us from the stated word of Darwin's arguments, relegate natural selection to a subordinate role, and attempt to add more meat (i.e. "other means of modification") to the bones. From the vantage point this creates, we can direct several questions to the *Origin*:

- *Origin of variations*: Either a variation is produced "by chance", and one then admits that, in Darwin's eyes, nature is the dominion of chance; or else, any random dimension that variation may have is denied and one posits that variation is governed by laws which must yet be discovered. What role, then, do external circumstances and the laws of internal development play with respect to the production of variations? Whatever response we give to this first question, it will presume nothing about the utility of variation.
- *Utility of variations*: Any variation, regardless of its cause, may be either useful or useless. It might, however, seem logical to say that variation has a higher probability of being useful if it has been produced by the environment and if there are laws in place to govern the organism's response to environmental changes.

In contrast to the figure of Darwin-the-Epicurean stands the figure of Darwin-the-Teleologist. Such interpretations of the *Origin* leave the random dimension of variation completely to one side, instead insisting on the principle of utility which *ultimately* governs the Darwinian process of evolution. As an example, Albert von Kölliker tends to deny chance its role, placing Darwin to the side of the teleologists—a view that is corrected by both Thomas Huxley and Carl Nägeli. A similar kind of reasoning is displayed by the Duke of Argyll, whose aim was to study "the reign of law". On the one hand, the Duke insists heavily on the way Darwin draws on ends-focused arguments, notably in his description of orchids; on the other hand, he also reminds us that chance is not a positive theory of Darwin's but (in keeping with Darwin's own terms) just a word cast over an ignorance. Both of these authors (Kölliker and Argyll) advance the image of Darwin-the-Teleologist and banish the concept of random variation from what can legitimately be taken for the core theory of the *Origin*.

Kölliker or the reign of utility

For Albert von Kölliker, there is no doubt that Darwin ranks among the teleologists.[13] This interpretation of Darwin's theory is, at the very least, odd, if not totally misguided and even incorrect. In a reply to Kölliker from 1864, Huxley made this observation: "It is singular how differently one and the same book will impress different minds". While Huxley reads the *Origin* as a complete rebuttal of Paleyite teleology, Kölliker understands Darwin as the very essence of teleology. Huxley generally praised the works of the renowned Würzburg professor of zoology: "all that proceeds from the pen of that thoughtful and accomplished writer, worthy of the most careful consideration".[14] But in a letter to Darwin, he articulated his own puzzlement, contrasting the high scientific status of the German naturalist to the bad reputation of the head of the French *Académie des sciences*: "Flourens I could have believed anything of: but how a man of Köllikers real intelligence and ability could have so misunderstood the question is more than I can comprehend".[15] Present-day readers tend to share in Huxley's astonishment: how could anyone see Darwin as the teleologist par excellence? The standard view, very much Huxleyan in this sense, presents Darwin as a critic of classical teleology and, as a result, Kölliker is overlooked by most of the literature on Darwin.[16] However, Kölliker's text immediately met with the honour of a translation in the *Reader*.[17] Kölliker's statement is worth consideration.

In Kölliker's view, utility is the most fundamental principle in the Darwinian conception of the world, trumping the principle of random variation. Kölliker also attributes to Darwin the idea that there is a "tendency to give birth to useful varieties (*Varietäten*)". Kölliker, then, simply puts two and two together and comes up with the conclusion that Darwin sees every part of the organism as *useful*! In contrast to what he considers to be the "Darwinian" teleological view, Kölliker asserts the equal perfection of each organism: "each animal is sufficient for its purpose (*Zweck*), is perfect in its kind (*Art*), and needs no improvement (*Ausbildung*)". For Kölliker, Darwin is a teleologist because he proclaims that each detail of the organism is constructed for the best (*zum Besten*). This is at least how Kölliker reads the passage from Chapter VI of the *Origin* where Darwin discusses "the protest lately made by some naturalists, against the utilitarian doctrine that every detail of structure has been produced for the good of its possessor". Those naturalists, Darwin adds, "believe that very many structures have been created for beauty in the eyes of man, or for mere variety", before concluding that "[this] doctrine, if true, would be absolutely fatal to my theory.[18]"

Kölliker's reading shows how this page of the *Origin* puzzled readers of Darwin's works. Directly after it, Darwin adds that some structures may not be altogether useless (which would be fatal to his theory), but simply not of direct use to their possessors, which then leads him to the following conclusion:

> Hence every detail of structure in every living creature (making some little allowance for the direct action of physical conditions) may be viewed, either

as having been of special use to some ancestral form, or as being now of special use to the descendants of this form–either directly, or indirectly through the complex laws of growth.[19]

So Darwin's final word is that he does not accept uselessness in the organism's structure. Even if no "direct use" is obvious, he strongly supports the existence of "special use", past or present. This is what strikes Kölliker as teleological:

> Darwin, in the fullest sense of the word, is a teleologist. He states quite precisely that every detail in the construction of an animal has been constructed for its greater good and his approach to the whole series of animal forms is from this point of view only.[20]

Huxley also concluded that "perhaps the most remarkable service to the Philosophy of Biology rendered by Mr. Darwin is the reconciliation of Teleology and Morphology";[21] although he added an important caveat to this:

> But it is one thing to say, Darwinically, that every detail observed in an animal's structure is of use to it, or has been of use to its ancestors; and quite another to affirm, teleologically, that every detail of an animal's structure has been created for its benefit.[22]

At the same time, Kölliker subtly adds, the Darwinian postulate of all-encompassing utility relies on a conception of the organism's imperfection and on the necessity, therein, for it to be perfected. Kölliker, on the other hand, is content to affirm the organism's *relative sufficiency* to its conditions of life, which implies neither perfection nor perfecting. In this, he leans on an 1861 brochure penned by the Austrian ornithologist August von Pelzeln, where we can find the following:

> The foundation, supposed by Darwin, for the utility of successively arising modifications rests on a confusion between differences in creatures' characteristics and the suitability (*Eignung*) of these to their purpose. Each creature is, in its own manner and for its own functions, equally perfect, and what is perfect cannot become more perfect.[23]

Pelzeln and Kölliker believed that they had identified, in Darwin's writings, a concept of evolution as a progressive advancement from less perfect to more perfect (*Fortschritt von unvollkommeneren zum vollkommeneren*), a principle of increasing utility against which they deemed it sufficient to counter with a first principle; that of equal perfection between all creatures.

The teleological interpretation of Darwin's vision of nature is possible only as long as the dark side of his philosophy, viz., the extinction of all that is not preserved, remains hidden from sight. Both Pelzeln and Kölliker felt justified in rebuking certain fundamental principles of the Darwinian world view; there is no "battle for existence", even if it is obvious that, were it not for the sheer number of unfavourable factors, the earth would soon be overpopulated. Their criticism resides in a blind spot of Darwinian theory: if useful variance is preserved, then this counts against both harmful variance and simple maintenance of the status quo. Thus, in any case, the systematic selection of all useful variance can only orient the Darwinian schema towards "the best"—a perfecting or progression. Darwin put forward no explanation to account for the production of variations and seemed to attribute some utility (past or present) to every part of the body; can we describe these as being perfectly adapted to their functions? For Kölliker, if this really is the case, then Darwin can only be understood as a teleologist. Kölliker draws our attention to the narrowness of the relationship between utility and natural selection. Thomas Huxley criticised Kölliker for several misreadings of the *Origin*. Countering him, Huxley asserts that Darwin never supposed "tendencies for organisms to give birth to useful *varieties*" and re-affirms that Darwin disregards all notion of necessary development or perfection, admitting only random *variations*. Switching between the two voices reveals the gap that lies open between *varieties* and *variations*. Today, Huxley's reading has prevailed against Kölliker's, although Kölliker's argument has also been given new life by James G. Lennox as well as through the critical analysis of Michael Ruse.[24] Indeed, the interpretation of Darwin as a teleologist is not merely some bizarre thesis linked to Kölliker's flawed understanding of the *Origin*.

Interpreted in accordance with the principle of utility, natural selection will tend to continuously work to make each part of the organisation more economical. Henceforth, any part the individual does not use will constitute a useless expenditure, the loss of which would therefore, in fact, constitute a net gain. The individual will profit from not wasting its resources on the construction of some part that is devoid of utility. Natural selection would favour the economy of material and the adaptation of parts to modified conditions. Such an interpretation can find support in several passages from Chapter V of the *Origin*, where the different causes of variation are described as being subject to the action of natural selection. Darwin believes that

> natural selection will always succeed in the long run in reducing and saving every part of the organisation, as soon as it is rendered superfluous, without by any means causing some other part to be largely developed in a corresponding degree.[25]

But this passage was modified for the fourth edition, where it is stated that "changed habits of life" are superfluous. Finally, the sixth edition reads:

> Thus, as I believe, natural selection will tend in the long run to reduce any part of the organisation, as soon as it becomes, through changed habits, superfluous, without by any means causing some other part to be largely developed in a corresponding degree.[26]

These results could be reinterpreted in the light of Goethe's views of nature, where he states that "in order to spend on one side, nature is forced to economise on the other side". Wilhelm Roux's post-Darwinian work on the struggle for existence within the organism itself would later revisit such notions, advancing that if there is development of one part then, correlatively, there will be atrophy of some other part.[27]

Ultimately, it is natural selection's job to *carry out* the transformations that use or disuse can only *authorise*. It does act, but only subject to the indications provided regarding the utility of the parts. Even its efficiency sees itself significantly attenuated between the first edition, where it "succeeds" in reducing the parts, and the sixth edition, where, losing its status as independent agent, it only "tends to" reduce.

Argyll and the defence of beauty

Kölliker's objection is a tenacious one, one that cannot be simply brushed aside. We find the same connection between the utility of parts and the origin of variations in the Duke of Argyll's criticisms. In his *Reign of Law* (1867), he observes Darwin trying to come to terms with the magnificent arrangements of orchid flowers.[28] From this, he asserts that Darwin is more concerned with answering the question "What is the use of the various parts, or their relation to each other with reference to the purpose of the whole?" than with enquiring "How were those parts made, and out of what materials?" Darwin is more concerned with "the use, object, intention or purpose of the different parts of the plant" than with the inner workings of flower anatomy.[29] Darwin is suspected of being interested only in the use, intentions, and *purpose* of the organism's various parts, and consequently not paying due attention to the ongoing tinkering within the marvellous apparatus of the orchids. The same reasoning can be maintained using certain passages from the *Origin*. Thus, we find Argyll quoting both Darwin's confession that "our ignorance of the laws of variation is profound", as well as the passage where the role of chance is presented as the mere confession of ignorance.[30]

If natural selection does operate on *materials*, then what are they? For Argyll, they are just those changes that are useful in the struggle for existence, since all change that has no utility value goes unaccounted for in the theory:

> Strictly speaking, therefore, Mr. Darwin's theory is not a theory on the Origin of Species at all, but only a theory on the causes which lead to the relative success or failure of such new Forms as may be born into the world.[31]

These reflections on the role of utility within the economy of the Darwinian system bring to light that Darwinian variation is not just some anatomical modification; it unfolds within a general ecology of relations. It could happen that purely neutral variations (neither useful nor useless) be produced, but natural selection would not apply to them and they would, therefore, *from the theory's perspective*, be as though inexistent. In this, we re-encounter the basis for Kölliker's objection: that Darwin's world view is teleological because, ultimately, it accounts only for useful variations. For the mechanism of natural selection to be operational, the variation must necessarily involve some value and this value must lead to some adaptation. Thus, Darwin's oversight is twofold: not only does he not account for the origin of variations (natural selection doesn't *originate* anything), but even among the variations produced (originated through potential laws which he does not speak of), he pays attention only to those that are more or less useful.

Extending Darwin-the-Teleologist into Darwin-the-Physico-theologian

Whether seen as promoting or undermining teleology, Darwin was both praised and criticised (Table 5.3).

Geneva botanist Alphonse de Candolle (1806–1893), son of Augustin-Pyramus but favourable to the idea of evolution and to the "philosophical spirit" of Darwin, regretted his British colleague's use of expressions such as "their corollas have been increased for that special purpose", where he should have written, in plain language: "their corollas being increased in size, the consequence is, etc". Or, instead of "subserve any special end", "have any effect". And Candolle adds:

> since observation only shows forms and consequences or effects, and not purposes or intentions. Our words "goal, end" ("*but, fin*"), suppose an intention, an external will. And, in order to know an intention, one has to question the one to which this intention has been attributed, or hear him speak his mind—events such as never occur in natural phenomena.[32]

Candolle aptly noted an ambiguity in the English lexicon: the words Purpose and End "have two contradictory meanings", suggesting either a "premeditated goal", or "an effect, a cause, a result". Probably, he surmised, translators did not pay enough attention to this difficulty and this may be the cause of "a confusion of ideas". Words entail a necessary vagueness, and this is why Candolle urged Darwin to avoid any expression that might suggest the supposition of intentions in nature, if he was going to pursue causes and effects in nature methodically.

Italian botanist Federico Delpino (1833–1905) is another interesting case of a naturalist who debated with Darwin without abandoning finalism. Delpino's programme was twofold: to enlarge the study initiated by Darwin in the *Orchids* and to amend the Darwinian theory of the *Origin* in light of teleology.[33] Delpino took teleology to be a simple way of fighting against the invasive system of materialism.

Clémence Royer, a strong opponent to Christian clerics, was nonetheless an "ardent deist"—a point well perceived by Darwin himself.[34] According to her

Table 5.3 Contrasted readings of teleology and physico-theology in Darwin

	For promoting teleology	*For undermining teleology*	*For having founded a new physico-theology*
Darwin praised	Asa Gray Federico Delpino	Thomas H. Huxley Ernst Haeckel	Clémence Royer (laws of progress) Al. Grant (it is unsatisfactorily mixed with an Epicurean framework)
Darwin criticised	Alphonse de Candolle Albert von Kölliker (Darwin puts too much emphasis on the principle of utility)	Karl von Baer	Argyll (Darwin mocked for not acknowledging it)

Source: adapted from Beatty 1990, p. 124 and Hoquet 2010a.

interpretation, Darwin rebuked the traditional arguments of the teleologists; but at the same time, she claims that Darwin founded a new kind of theology:

> Mr Darwin's book may be, among those I've read, the one which gives the strongest incentive to believe in God, the only book which succeeds in apologising for the world as it is; it is an eloquent theodicy in action, which leaves far behind it all former attempts of theologians and those rhetorical philosophers that Voltaire (to whom everything was permitted in matters of language) called the final-causers (*cause-finaliers*).[35]

While rejecting finalism, Darwin proposes a new kind of justification of the world; a new explanation for the harmonies and beauties of the natural world, explained by law—but also an explanation for nature's defects and blunders. Royer claims that the *Origin* is the basis for a new kind of natural theology which also accounts for dysteleological features. Darwin's system clearly evinces the fact that nature is made up of both perfection and imperfection mixed together—all unplanned effects of blind mechanical or natural causes. The absence of consciously designed contrivances in Darwin's *Origin* does not entail the end of natural theology; the overall view of nature remains ultimately redemptive.

Other readers of Darwin also came away with a teleological reading of the *Origin*. In the marginalia to Royer's translation, French palaeontologist Albert Gaudry notes:

> Everything that is said on natural selection [in Royer's translation: *élection naturelle*] proves the direct intervention of the creator [words crossed out by Gaudry] of God.... Without this intervention, all [those pages] have no meaning.... Reading this book proves God's continual action.[36]

Similarly, the American Alexander Grant considered Darwin as both an Epicurean philosopher and a new natural theologian: "the [Darwinian] theory is the theory of Epicurus, with the atheism removed", since "there is nothing atheistical in Mr. Darwin's work; on the contrary, it might be described as a system of natural theology founded on a new basis".[37] It seems that, for many readers, Darwinian teleology is a stumbling block: once teleology is maintained in the form of laws (such as natural selection), it becomes impossible to get rid of Divine Providence.

For this reason, teleological readings of Darwin insist on two aspects of natural phenomena. Kölliker insists on the utility of organs, a necessary product of the Darwinian system; Argyll's emphasis is on the overarching and irrepressible presence of beauty and an aesthetic sense, both liable to be preserved even where they may be useless. Such readings are an invitation for us to rethink our strategy when answering two of the *Origin*'s most ponderous questions: does Darwin support a utilitarian re-interpretation of the entire organism? And, is Darwin's nature ruled by an all-encompassing mechanism, or does the existence of natural selection leave room for aesthetic criteria which escape natural selection's dividing blade?

Darwin's emphasis on "wonderful contrivances" brings teleology back into nature with force, allowing Asa Gray to claim that "natural selection is not inconsistent with natural theology".[38] Darwin is a kind of teleologist as he acknowledges design and contrivances in nature, and also because his interpretation of organic forms leans heavily on a principle of usefulness. In the *Origin*, Darwin's enemy is not so much the Bridgewater treatises as special creationism. Darwin opposes his view to special creations, but his theory of the modification of species by laws of descent is susceptible to being interpreted as new terrain for physico-theology, as C. Royer, A. Grant, and A. Gray all suggested. All claimed the existence of a physico-theology that was truly Darwinian, even though Darwin had no intention to produce any such thing. Physico-theology is not synonymous with the theory of special creations. It refers only to any kind of inference from natural facts to the Creator; it is a theological discourse grounded in scientific results (whatever they may be). In the kind of natural theology attributed to Darwin, God had no foreknowledge of the particular forms that life would take: everything is operated by the designed laws of nature, not by brute force and the proof of God's existence can be based on the apparent perfection resulting from the blind, cruel, severe, yet lawful and designed, elimination of imperfections.

By broaching "contrivances", especially in his *Orchids*, Darwin set himself on a slippery slope that would inescapably veer from teleology to theology. The two questions should nonetheless be treated distinctly. Teleology deals with utility and adaptation in organisms and theological assumptions traditionally come attached to it. But even without teleology in the classical sense of design (intended order and perfect adaptation of means to ends), the Darwinian framework served as the grounding for a new kind of natural theology where the selection of useful structures was understood as a *law of progress*.

Utility, selection, perfection

The various readings we have seen should not be rejected as mere "misreadings". They are all perfectly grounded in Darwin's text, as Alphonse de Candolle noted in his letters to the great man himself. The reason Darwin can be bent equally well to both Epicurean and teleological interpretations is that his text presents a certain number of ambiguities. In the *Origin*, Chapter III, for instance, begins with deep Paleyan accents, referring to "all those exquisite adaptations of one part of the organisation to another part, and to the conditions of life, and of one distinct organic being to another being … these beautiful co-adaptations".[39] Judged on this, one must presume that his text provides an anti-teleological explanation for such sophistication and beauty. But at the same time, throughout Chapter III, Darwin insists upon the existence of laws and sets out to definitively refute chance:

> Throw up a handful of feathers, and all must fall to the ground according to definite laws; but how simple is this problem compared to the action and

reaction of the innumerable plants and animals which have determined, in the course of centuries, the proportional numbers and kinds of trees now growing on the old Indian ruins![40]

A *falling* feather seems simpler to explain than the *form or formation* of a feather; the laws of nature are not all of equivalent simplicity.

By establishing the struggle for existence as a general theory of "the mutual relations of all organic beings",[41] Darwin shows that this is the true object of his search: simple relations. Although, in the process, what is "exquisite" is reinterpreted through the theory of natural selection, the chapter does not end before evoking a consoling vision which seems to suggest that there is progress, some orientation in evolution towards what is better, what is of higher utility. So, while the reinterpretation of beauty on the basis of blind mechanisms argues in favour of Epicureanism, the existence of laws of progress, by contrast, grounds the idea of a teleological Darwin.

Chapter IV entirely extends these same lines, once again evoking "the beauty and infinite complexity of the coadaptations between all organic beings".[42] The vocabulary employed here is awash with words like *better* ("infinitely *better* adapted", "*better* adapted forms", a place "*better* filled by some modification"), *improved* (modified and *improved*; *to improve* still further; *highly improved*), *perfected* (modified and *perfected*); all culminating in, "nature's productions ... should be infinitely better adapted to the most complex conditions of life, and should plainly bear the stamp of far higher workmanship".[43] Natural selection always preserves what is "advantageous", "favourable", "profitable": on the other hand, "we may feel sure that any variation in the least degree injurious would be rigidly destroyed".[44] Relating this back to the diagram from Chapter IV, we would say that, yes, at some time t, all creatures occupy the same horizontal, all are equally adapted; but if we then compare the creatures from time t with those from time $t-1$, then a gap or distance created by a process of perfecting becomes apparent.

Darwin even speaks of a "general law of good being" present in nature, resulting, for example, from the necessary intercrossing of individuals.[45] In several passages of Chapter IV he mentions "the natural economy", the "polity" of nature, i.e. its rational and infinitely wise order.[46] He underlines the manner in which natural selection modifies organisms in order that "we should then have places in the economy of nature which would assuredly be better filled up": "natural selection will always tend to preserve all the individuals varying in the right direction, though in different degrees, so as better to fill up the unoccupied place".[47] Here we see a direction in variation and an efficiency in the laws governing nature, both guaranteed by natural selection, both inscribed at the heart of Chapter IV.

The issue around the utility of characters is just as widely debated. In an addition to the fifth edition, Darwin seems to extend the utility of organs (and, thereby, the dominion of natural selection) when he states:

> No one will maintain that we as yet know the uses of all the parts of any one plant, or the functions of each cell in any one organ. Five or six years ago,

endless peculiarities of structure in the flowers of orchids, great ridges and crests, and the relative positions of the various parts would have been considered as useless morphological differences; but now we know that they are of great service, and must have been under the dominion of natural selection.[48]

But this passage is wiped from the sixth edition and replaced by a passage in Chapter VII where Darwin confronts the objection that "many characters appear to be of no service whatever to their possessors, and therefore cannot have been influenced through natural selection".[49] Darwin acknowledges the strength of the objection and deploys a full arsenal of responses to it: prudence is no longer (as it was in the fifth edition) a matter of abstaining from the conclusion that certain parts are *useless*, but, quite to the contrary, a matter of no longer hastily forming judgements of *utility*; laws of correlation, and "laws of growth" in general, create a mutual pressure on development and fulfil a direct action of the conditions of life. Darwin seems here to be leaning further and further towards the idea that all structures have utility.

We have already analysed how another passage from Chapter VI is just as much a source of ambiguity and dispute between Kölliker and Huxley.[50] In this passage, Darwin ultimately admits there no seeming uselessness and, though he may acknowledge the apparent absence of *direct use*, this is only in order to better redeploy utility at different levels of *special use*, present or past. Thus, here, Darwin is clearly stating that all structures are governed by a principle of utility.

Responding to this, Huxley endeavours to defend Darwin by showing how Kölliker had confused the simple observation of organ utility with the teleological interpretation whereby each organ has been specially created with a view to assuring some certain function, and Huxley himself, on the other hand, calls for the two dimensions to be clearly dissociated.[51] As for the idea that, as a result of this attention to utility, Darwinian variations must be "orientated", Huxley can only point out that Darwin, to his knowledge, never wrote such a thing. In all this, he flatly rejects Kölliker's interpretation, which is based on a certain relation between natural selection and utility of parts. The *non-purposive* character of Darwinian variation is clearly established, notably in the last sentence of Chapter V where it is stated that, "whatever the cause may be of each slight difference in the offspring from their parents—and a cause for each must exist", natural selection accumulates "such differences, when beneficial to the individual".[52] But, in this, Darwin establishes a very close link between natural selection and utility—and this is why his system is sometimes interpreted as a teleological system or a system of progress.

Uselessness, harmfulness, beauty

Darwin does, however, try to embrace uselessness within his view, or at least provisionally; the idea of laws of correlation between variations allows for the reach of nature's lawfulness to be extended without having to subject everything to the dominion of natural selection. Above all, it allows us to explain the formation of useless organs (or organs whose utility is provisionally undetermined). Darwin seems to have accepted one exception to the rule of organ utility: correlation between characteristics dispenses with the obligation for each of them to be directly useful. No concession, however, is extended for the case of "free" beauty.

Darwin himself evokes the existence of utterly useless organs in the recap chapter to the *Descent of Man*. These are

> structures, which as far as we can judge with our little knowledge, are not now of any service to [Man], nor have been so during any former period of his existence, either in relation to his general conditions of life, or of one sex to the other. Such structures cannot be accounted for by any form of selection, or by the inherited effects of the use and disuse of parts.[53]

The "correlation of growth" idea allows Darwin to include characters of the organism which appear to be useless; in doing this, he favours internal laws over the external environment, laws of variation over the efficiency of natural selection.

The utility of a structure or character is, moreover, open to debate. First, an organism is not to be considered in isolation but only in relation with other individuals (sexual partners specifically). Second, organs are also not to be considered in isolation, but only in relation to other organs within a general structure. In both cases, it seems that we do not possess the necessary criteria to guarantee a measure of the usefulness or uselessness of a given character.[54] In that respect, the problem Eduard von Hartmann sees in co-adaptations is that utility does not have its end within a single organ, rather its domain is found to be somewhat splayed out. In the case of a flower having sweet sap at the bottom of a long calyx and an insect being endowed with a long horn for sucking,

> none [of these parts] is useful in and of itself; it has value only within the hypothesis of correlative propriety; hence, neither [part] offers an advantage in the struggle for existence unless the corresponding disposition of the other part is supposed as already granted.... We are thus forced to admit the rigorously parallel advancement of both modifications.[55]

Here, Darwin dives straight into the abyss of the unthinkable, endeavouring to get around the objection of uselessness, even harmfulness: he appeals for prudence in declarations of utility and uselessness, recalling that nature, more insightful than us, can identify an advantage where the human eye may discern nothing at all; he also indicates, in contrast, that the "utility" of a character is not measured only in terms of the immediate benefit to its individual possessor, but may constitute some reproductive advantage (i.e. that the individual will produce a greater number of descendants). A verdict of uselessness is therefore always suspect, since utility can be hidden or relative. In other words, we cannot identify what does or does not contribute to "fitness".

Finally, the question of "useless" characteristics comes down to the objection of beauty and its place in nature. If beauty has no utility for the individual, then it can find no explanation within natural selection, it's just something left over in the organism, awaiting some natural reason. If, however, beauty is harmful to the individual, then its existence actually contravenes natural selection. In both cases, the idea of beauty existing in nature independently of any utilitarian consideration may be seen as an opening for Providence and physical theology.

In order to include the set of characteristics judged to be "beautiful", Darwin extends the domain of utility through the concept of "sexual selection".[56] In an example from the section "Organs of little apparent importance" in Chapter VI of the *Origin*, Darwin reveals that two competing explanations can be given for the woodpecker's green colour. The first explanation, which takes it for an adaptation to life among leaves, would be valid if woodpeckers with plumage of other colours did not exist. The second explanation, the one Darwin holds to, connects the green colour to sexual selection.[57]

The Duke of Argyll formulated the objection to uselessness with the most vigour: the colours of animals could well be explained by a principle of beauty and harmony at work throughout creation. He expressed shock at Darwin's opinion that the admission of beauty's place within nature would be fatal to his theory.[58] Hadn't Darwin, at the very outset of his reflections, spoken of the "beautiful co-adaptations" that characterise the relationships between living forms?

For Darwin's opponents, the phenomenon of co-adaptations is an invitation to replace the struggle for existence with a regulated process, a correlative law of

Table 5.4 Beauty and natural selection

	Relation to natural selection	*What the* Origin *says*
Beauty as useless characters	Unexplained by natural selection	Utility is more common than expected: natural selection is more insightful than artificial selection
		Utility can be derived or indirect, thanks to the operation of laws of correlation of variations
Beauty as harmful characters	Contradictory with natural selection	Natural selection is completed by sexual selection
		Selection is not only a dividing blade but also involves access to copulation

evolution. Opponents of this bent suppose a general process of ideal harmony, whose correlation laws, operating between the different parts of individuals, are in themselves just expressions. In any case, the struggle for existence is not entirely excluded, rather it is just pushed back to the rank of auxiliary principle.[59]

Progress from the first rudiments

Another form of teleology is mixed into Darwin's texts through the concept of some "direction", if not to say "progress", to evolution. This aspect was very clearly put forward by Darwin's first French translator Clémence Royer, who interpreted the *Origin* as an exposé of the "laws of progress". Such a reading directly contradicts the extremely clear position that Darwin had laid out in an important passage of Chapter IV where he refutes progressive development.[60] But one can just as easily draw differing conclusions from other passages of the *Origin* where Darwin mentions that "we have reason to believe that such low beings change or become modified less quickly than the high".[61] The fact that he does not avoid the opposition between "high" and "low" organisms indicates an admission of hierarchy between the different forms of life, something which, problematically, opens out onto the idea of "progress" in nature.

Many readers of the *Origin*, whether they believe themselves to be supporting or refuting it, orient themselves towards the laws of variation to the detriment of natural selection which is in each instance distanced, disqualified, or downplayed. The value of the *two-step process* is contested and all interest is shifted to the production of variations, the indispensable material upon which natural selection operates. Regardless of whether Darwin supports a theory where it is blind chance or inflexible laws which produce variations, in either case natural selection has no power to produce them: it risks becoming nothing more than a destructive agent, creating nothing. Its powers would be limited to sanction and severance, but as for the origin of variations (the inherently creative or originating phase of the process), here it would be entirely impotent.

To the standard image of a selectionist Darwin we have seen the contrasted figures of Darwin-the-Epicurean and Darwin-the-Teleologist. Exploring this here involves no claim to speaking the truth about "Darwinism", nor any attempt to crumble its unity into a multitude of perspectives. It is simply a matter of acknowledging that different readings of Darwin were indeed explored, readings whose pertinence can be contested, but which inarguably focus our attention on certain, specific aspects of Darwin's text, notably the ambiguous place accorded to utility, to progress, and to beauty within his system. Further still, some of these readings even claimed allegiance to Darwin's very own thinking, all the while radically contesting certain fundamental postulates of the classical two-step presentation of Darwinism. In reading Eimer, Berg, Kölliker, Argyll, or Royer and Gray, we see just how plausible it is to understand the arguments of randomness and utility as being incompatible and to then deploy them one against the other.

Notes

1 *Origin* 1859, p. 171, *Var* 321 (#5); 1859, p. 199, *Var* 367 (#208).
2 *Origin* 1872, p. 158, *Var* 364 (#191:f).
3 Darwin 1868, vol. 2, pp. 248–249.
4 *Origin* 1859, p. 43, *Var* 119 (#322). See also Darwin to Hooker, 23 November 1853, CCD 6 282.
5 Nägeli 1865, p. 16 and note pp. 17–18.
6 Huxley 1893, vol. 2, p. 87.
7 Weismann 1868, p. 18.
8 Mayr 1962, p. 5.
9 Sober 2000, p. 37.
10 Weismann 1868, p. 20.
11 Mayr 1961, p. 365; Ruse 2003, p. 122.
12 Ruse 2013, p. 12 and the chapters by Beatty and Lennox in Ruse 2013, pp. 146–157.
13 Kölliker 1864, p. 175 and p. 178.
14 Huxley 1893, vol. 2, p. 81.
15 Huxley to Darwin, 5 October 1864, CCD 12 346.
16 The detailed works of Peter J. Bowler (1983, 1988) on alternative non-Darwinian theories of evolution do not mention Kölliker. Montgomery 1972 (p. 86) cites Kölliker among those writers who have accepted evolution but were not Darwinians. Kölliker is a puzzle to Montgomery since his scientific field (invertebrate zoology) contains a surprisingly large proportion of Darwinians: thus, Kölliker stands out as a sort of exception in his own discipline. Duchesneau 1987 (pp. 233–253) analyses Kölliker's place in the field of histology.
17 *The Reader*, 4 (13 August 1864), pp. 199–200; (20 August 1864), pp. 234–235.
18 *Origin* 1859, p. 199, *Var* 367 (#206–208). In [d], this passage would become a separate section: "Utilitarian doctrine how far true: beauty how acquired".
19 *Origin* 1859, p. 200, *Var* 369 (#220).
20 Kölliker 1864, p. 175.
21 Huxley 1887, vol. 2, p. 201.
22 Huxley 1893, vol. 2, p. 87.
23 Pelzeln 1861, p. 7.
24 Lennox 1993 and 1994; Ruse 2003.
25 *Origin* 1859, p. 148, *Var* 296–297 (#136).
26 *Origin* 1872, p. 118, *Var* 297(#136:f).
27 Goethe is cited in *Origin* 1859, p. 147, *Var* 295; see also DNS, p. 304. Roux 1881.
28 Darwin 1862.
29 Argyll 1867, pp. 39–40. Contrast this reading with Asa Gray's famous claim that the *Orchids* are "a beautiful flank movement". On this, see Lennox 1993 and 1994, Ghiselin 1994, Hoquet 2010, and Tabb 2016.
30 *Origin* 1859, p. 167 and 131; *Var p.* 317 and 275.
31 Argyll 1867, p. 219.
32 Alphonse de Candolle to Darwin, 31 July 1877, in Baehni 1955, pp. 147–148.
33 See Pancaldi 1984.
34 Darwin to Asa Gray, 10–20 June 1862, CCD 10 241.
35 Royer 1862, pp. xxxiii–xxxiv.
36 Tassy 2006, p. 49.
37 Grant 1871, pp. 275, 281.
38 Gray 1877, pp. 72–145.
39 *Origin* 1859, p. 60, *Var* 144 (#8–9).
40 *Origin* 1859, p. 75, *Var* 158 (#130).
41 *Origin* 1859, p. 78, and IV, p. 80, *Var* 162 (#163) and 164 (#9).
42 *Origin* 1859, p. 109, *Var* 202 (#228).

43 *Origin* 1859, p. 84, *Var* 168 (#39).
44 *Origin* 1859, p. 81, *Var* 164 (#12).
45 *Origin* 1859, p. 99.
46 *Origin* 1859, p. 102, 104, 108, 122; VI, p. 173, 178.
47 *Origin* 1859, p. 81 and p. 102, *Var* 166 (#20) and 193 (#178).
48 *Origin* 1869, p. 152, *Var* 234 (#382.65.0.10–11:e).
49 *Origin* 1872, p. 170, *Var* 232 (#382.62.x-y:f).
50 *Origin* 1859, p. 200, *Var* 369 (#220).
51 Huxley 1893, vol. 2, pp. 82–87.
52 *Origin* 1859, p. 170, *Var* 320 (#305).
53 Darwin 1871, vol. 1, p. 387.
54 Cf. Darwin to Charles Kingsley, 10 June 1867, CCD 15 298.
55 Hartmann 1877, p. 79.
56 On Darwin's debates related to sexual selection, especially with A.R. Wallace, see the review by Hoquet and Levandowsky 2015.
57 *Origin* 1859, p. 197, *Var* 364 (#193).
58 Argyll 1867, p. 183 and p. 197, respectively.
59 On auxiliary principles, see the section devoted to Hartmann, Chapter 6 and Table 6.2.
60 *Origin* 1861, p. 135, *Var* 223 (#382.18–20), text introduced in (c) and partially revisited in (e).
61 *Origin* 1859, p. 133 and 388, *Var* 278–279 (#29) and 617 (#42).

6 Darwin-the-Lamarckian and the other "means of modification"

Giving Chapter V its place

Chapter V, entitled "Laws of variation", has not grabbed the attention of readers as it might. Coming directly after the sequence whose centrepiece is the presentation of natural selection and whose finale is Chapter IV's image of "ever branching and beautiful ramifications", its fate is too often overlooked.[1] This general neglect of Chapter V might be explained by a number of reasons. Either readers don't know what place to give it within the economy of the book, or else the themes it discusses seem precariously Lamarckian.[2] But, whatever the reason, a general silence does seem to hang over its content, seemingly foreign to the definition of "Darwinism", and this despite its pivotal position within the *Origin*, just prior to Darwin's examination of the difficulties his theory meets.

Darwin's theory leans on the existence of variations, defined as "mere individual differences". Darwin affirms the fundamental plasticity of individuals, and *plastic*, in his lexicon, is a synonym of undefined variant. It is of the utmost importance to his theory that variations result from *chance* or *accidental variations*, as opposed to variations produced along determined lines (e.g. immediate adaptations to environmental demands). And yet, Darwin opens Chapter V with the affirmation that to be "due to chance" "is a wholly incorrect expression, but it serves to acknowledge plainly our ignorance of the cause of each particular variation". Thus, the purpose of Chapter V is to affirm that variation is not produced "by chance" but rather in accordance with laws which we do not yet know, but must endeavour to know.

Later, in Chapter XIV ("Recapitulation and Conclusion"), Darwin evokes "a grand and almost untrodden field of inquiry ... on the causes and laws of variation, on correlation of growth, on the effects of use and disuse, on the direct action of external conditions, and so forth".[3] This phrase, found in the text of the *Origin* itself, grants the reader carte-blanche by authorising hypotheses of all sorts for apprehending the laws of variation and inheritance (even including— heaven forbid!—the heredity of acquired characteristics). Hence, with respect to specifying these laws, themes normally associated with non-Darwinian modes of transformism rear their heads again here. Chapter V's introductory list of sections may be surprising in its evocation of themes with quite a Lamarckian flavour: "Effects of external conditions—Use and disuse, combined with natural selection; organs of flight and of vision—Acclimatisation"; or maybe because it suggests a variation that would no longer be random and undetermined, but rather directed according to certain "tendencies", like in the section entitled "Species of the same genus vary in an analogous manner". The massive presence of these Lamarckian or orthogenetic themes in Chapter V explains why today's Darwinians brush it aside, as though it were purely incidental to the economy of the book; as though Darwin allowing room for Lamarckian mechanisms was a kind of afterthought and that they could easily be removed from the infrastructure of the system without bringing it tumbling down, since the real agency in Darwin's theory belongs not to the laws of variation but to natural selection.

In keeping with its title, Chapter V's aim is to provide some of the laws of variation. At several points, Darwin states and then restates that he has significant facts at his disposal but that, unfortunately, the present work is not at all the place to reveal them. Among these "laws", various so-called "Lamarckian" factors are studied. Most often, their direct role is reduced to a minimum, with all the most important effects being ascribed to natural selection. What Darwin calls "Correlation of growth" is also connected to natural selection with a principle relating to a general economy of resources. Darwin does his best to give several rules of variation, notably the rule stating that, "*A part developed in any species in an extraordinary degree or manner, in comparison with the same part in allied species, tends to be highly variable*".[4] Although repeatedly insisting that this proposition is "a rule of high generality", Darwin never sticks to his word on the matter and instead constantly returns to how natural selection can explain it.

That Chapter V is subordinate to the themes of Chapter IV (i.e. that the laws of variation are subordinate to natural selection) is unquestionable. Each and every theme called upon in Chapter V is systematically offset by the action of natural selection. When Darwin questions the respective roles of natural selection and the conditions of life, he concludes that the latter play only an indirect role (through their effects on the reproductive system, i.e. as a cause of variability) whereas the accumulative action of natural selection is the real manufacturer of characteristics.[5] Similarly, cases of use and disuse (wingless or blind animals) and laws of correlation of growth are connected to the action of natural selection.

Nevertheless, Chapter V is unique in its emphasis on variation and the laws of variation. With respect to each of these laws, Darwin constantly reaffirms the fact that descent with modification (the theory of the unknown common ancestor) proposes a "*vera causa*", in contrast to "the ordinary view of each species having been independently created", a view which must presume imaginary or unknown causes.[6] Darwin especially develops these points in relation to the bluish colour of pigeon plumage, though he also mentions the stripes found in the equine family, something he sees as a primitive characteristic: "not even a stripe of colour appears from what would commonly be called an accident".[7] He adds that descent with modification is linked to the emergence of a "*generative variability*" and that both can be produced through either sexual or asexual reproduction.[8]

The decisive importance of variation within the mechanism of natural selection is further indicated when Chapter V successively reframes the main themes from Chapter I as dealing not with artificial selection but with the manner in which a change in the conditions of life produces a functional disturbance in the reproductive system of the parents, consequently producing "the varying or plastic condition of the offspring"; as establishing that "use in our domestic animals strengthens and enlarges certain parts, and disuse diminishes them; and that such modifications are inherited".[9]

On top of this, Chapter V frequently evokes the pigeon theme and, accordingly, can be seen as a kind of "Chapter I (Reprise)". With its strong support from Chapters I and IV, Chapter V demonstrates its rightful place within the general economy of the *Origin*. Nevertheless, a large number of Darwin's initial readers singled it out as their bone of contention with the work. The argument goes like this: Darwin seems to think that "natural selection acts solely by the preservation of profitable modifications";[10] it therefore depends entirely on the material upon which it works, yet this material is provided by variation. This being the case, it is variation and its characteristics which drive the system's deployment. Natural selection is capable of everything, but only once variations have been produced and this is why the laws of variation are logically and chronologically primary. Several times, Darwin even admits that variation is the necessary condition for the agency of natural selection. So, is there contradiction between selection and the laws of variation? Does the *Origin* offer us a satisfactory response on this point? These are the two wedges that readers have tried to drive into the Darwinian project and which Chapter V must consequently take on. What if Darwin's hypothesis of transformation by means of natural selection was secretly undermined by his commitment to the laws of variation, to use and disuse, and by his belief in the overarching importance of habit?

Traces of Lamarckism can be infused into the very heart of Darwin's system, as, for instance, when his German translator, H.G. Bronn, initially chose to render *natural selection* as "*die Wahl der Lebens-Weise*" (or the choice of a way of life).[11] In the face of competition, certain individuals are forced to adopt modifications to their lifestyles, making this newly chosen "way of life" "the most fruitful and general cause of production of varieties". For Bronn, the overabundance of offspring leads not to *eliminations* but to *reconversions*, not to *suppressions* but to *reorientations*. Instead of struggling for the same habitats or resources, individuals expand the range of those habitats and resources traditionally exploited by their ancestors. By interpreting this "choice of lifestyle" as a form of sympatric speciation through divergence of characteristics and habits, Bronn's expression falls in line with an idea Darwin denoted using an expression borrowed from Milne-Edwards: "division of labour", a principle permitting improved allocation of a region's resources, in turn enabling the region to accommodate a larger number of forms. Bronn's was therefore a constructive interpretation of natural selection, where competition pushes individuals along the path of alternative habits, and in this he avoided the risk of interpreting *natural selection* as a crude dividing blade; his interpretation indicates how nature invents by working on the margins.

But Bronn's choice can also be interpreted in a different manner. Now, rather than the (primary) modification prompting the choice in lifestyle, it is the choice in lifestyle that modifies the organism. Understood in this way, Darwinian *natural selection* becomes a concept of use and disuse. Like Lamarck's "habits", Bronn's *Wahl der Lebens-Weise* would explain differentiation of forms through the continuous action of differences in lifestyle, with these in turn leading to differences between the faculties. Darwin, of course, was adamantly averse to this translation, whose Lamarckian accents he denounced in no uncertain terms. Inquiring into other possible equivalents for his "natural selection" (beginning with *Adelung*, "ennobling, would or perhaps be too metaphorical"[12]), he shifted the search towards the vocabulary of breeders. But on the case of Bronn's initial translation of *natural selection*, we see how easily Darwin's thinking can be tainted by Lamarckian overtones.

"The main but not exclusive means of modification"

If *descent* refers to the different ways in which individuals create lineages, according to ascendant/descendant relations, what about the *means of modification*? Is natural selection the full extent of Darwin's response to this question? Undoubtedly one of Darwin's most quoted sentences on that matter is the following: "Furthermore, I am convinced that Natural Selection has been the main but not exclusive means of modification".[13]

This statement, illuminating the "by means of natural selection" from the book's title, is not found hidden in the correspondence; it is not a confession dragged out of Darwin, neither is it a concession ceded in some later edition of one of his books in order to satisfy his detractors. It is situated, in plain sight, right at the beginning of the *Origin*, just as the Introduction comes to a close. The passage was present in 1859 and remained there through all subsequent editions, undergoing only the most minor of changes, and finally being augmented by a long comment in the sixth edition where Darwin bemoans precisely the fact that no one has paid any attention to his warnings.[14] After more than ten years of debate around the *Origin*, Darwin judged it necessary to revisit the statement, this time hammering his message out unequivocally and with not just a little exasperation.[15]

Here we see Darwin insisting on the fact that his readers have persisted in refusing to see one point he had underlined: by acknowledging the existence of a gap between natural selection as *main factor* and natural selection as *exclusive factor* in the modification of species, he had intentionally allowed room for other mechanisms. Now he is led to denounce this widespread error of interpretation which totally ignores his cautions. He also communicates his hope that the foreseeable future will see the error corrected. The naturalist mechanism of natural selection does constitute an essential grounding point of what Darwinism necessarily evokes. Darwin himself, however, seemed eager to get beyond an overly narrow selectionism.

Many other of Darwin's texts reinforce the lesson to be drawn from these earlier extracts. In *The Descent of Man* (1871) he declares, by way of repentance: "I probably attributed too much to the action of natural selection or the survival of the fittest", before clarifying that he

> had two distinct objects in view, firstly, to shew that species had not been separately created, and secondly, that natural selection had been the chief agent of change, though largely aided by the inherited effects of habit, and slightly by the direct action of the surrounding conditions.[16]

Such passages have been given quite contradictory interpretations. Even among recent historians and philosophers, similar disagreement can be found. To take three emblematic examples, James Moore thinks Darwin's theory still stood in the sixth edition, though "neither so elegantly nor impressively as before";[17] Robert Young suggests a "useful exaggeration" that the book should have been re-titled "on the origin of species by means of natural selection and all sorts of other things".[18] As for Jean Gayon, he cites both passages from the 1859 introduction and the addition to the sixth edition, as *supporting* Darwin's belief in the paramount power of natural selection. In Gayon's view, when in the final chapter of the last edition of the *Origin of Species* (1872) Darwin "states that he had not changed his opinion as to the respective roles of natural selection and of other factors involved in the modification of species, repeating the formulation used in the introduction to the first edition", this means that he was reiterating his support of natural selection.[19]

But other readers have identified the same words as displays of Darwin's remorse; a stack of proof establishing that Darwin did indeed regret every turn of phrase he may have employed that tended to give exclusive privilege to natural selection.[20] In this framework, these passages become so many weapons for those dubious of the magnitude of natural selection's role; weapons forged from Darwin's own words, thus ably equipping them to pit Darwin against Darwin. Or, more specifically, to pit Darwin against the authority of the pure or radical selectionists claiming to be the sole heirs to the Master, notably Wallace and Weismann. Countering derivations that redefined Darwinism as pure selectionism, the Darwin of 1872 looked to later developments in science hoping they might definitively resolve the issue. A certain number of his savant readers, such as Alphonse de Candolle, also appealed for others to "distinguish the theory of the derivation of forms from the necessary fact of the selection of these forms once they have been produced".[21]

Conversely, such deviations of Darwinian doctrine, despite their source in Darwin's own words, were totally rejected by the radical Darwinians, who declared that this was blatant "Lamarckisation" at work, if not actually Darwin's own still latent "self-Lamarckisation". The role the term "Lamarckism" played in this debate was as the correlate and logical opposite of "pure Darwinism". It is the question mark over various factors (those irreducible to natural selection; use and disuse, influence of exterior circumstances) and their place within the theoretical structure of Darwinism.

Darwinism or Wallacism?

Staying on the topic of natural selection, there are two ways to read Darwin's words when he says that it is not "exclusive": either we integrate these passages as a component in our understanding of Darwinism, and, in doing so, remain loyal to Darwin's word; or else we take them to be external to the logic of the Darwinian system and reject them along with everything else that is not selectionist.

The second reading corresponds to Wallace's strategy of purification (and absorption) of Darwinism, the paroxysm of which was the publication of his book *Darwinism* (1889). This title forthrightly announces its intention to return to the very essence of Darwin's lesson, as it was before it became deformed under the hammer blows of the various criticisms brought down on it. Wallace explains this in the subtitle to his book: to speak of Darwinism is to give "an exposition of the theory of natural selection with some of its applications". Such a declaration can certainly find footholds in Darwin's own words. For example, when the latter explains that

> it long remained to me an inexplicable problem how the necessary degree of modification could have been effected, and it would have thus remained for ever, had I not studied domestic productions, and thus acquired a just idea of the power of Selection.[22]

But Wallace drives the point in deeper, interpreting Darwin in the direction of a pure selectionism which, in fact, far overshoots his stated position:

> Whatever other causes have been at work, Natural Selection is supreme, *to an extent which even Darwin himself hesitated to claim for it*. The more we study it the more we are convinced of its overpowering importance, and the more confidently we claim, in Darwin's own words, that it "has been the most important, but not the exclusive, means of modification".[23]

Wallace's proclaimed "Darwinism" is intended to carry Darwin to the logical conclusion of his own thought, a conclusion to which Darwin himself did not come and never would have asserted. From the end of the *Origin*'s introduction, Wallace keeps only the statement which makes natural selection the "main" agent. Wallace's intention was to return to a strict Darwinism that declares the *agency* of natural selection in the formation of the species loud and clear, to the exclusion of all other principles. In particular, "Mr. Darwin has shown that, in the distribution and modification of species, *the biological is of more importance than the physical environment*, the struggle with other organisms being often more severe than that with the forces of nature": on such a basis, making any room at all for the influence of the environment is simply ruled out.[24]

Conversely, this same statement from Darwin can lend itself to another interpretation entirely, simply by placing the accent on the second half of it: this

Table 6.1 "Wallaceism" and Darwinism compared according to G. Romanes (1892, vol. 2, p. 6)

The theory of natural selection according to Darwin	The theory of natural selection according to Wallace
Natural selection has been the main means of modification, not excepting the case of Man	Natural selection has been the sole means of modification, excepting in the case of Man
(a) Therefore, it is a question of evidence whether the Lamarckian factors have co-operated	(a) Therefore, it is antecedently impossible that the Lamarckian factors can have co-operated
(b) Neither all species, nor, *a fortiori*, all specific characters have been due to natural selection	(b) Not only all species, but all specific characters, must necessarily have been due to natural selection
(c) Thus, the principle of Utility is not of universal application, even where species are concerned	(c) Thus, the principle of Utility must necessarily be of universal application, where species are concerned
(d) Thus, also, the suggestion as to Sexual Selection, or any other supplementary cause of modification, may be entertained; and, as in the case of the Lamarckian factors, it is a question of evidence, whether, or how far, they have co-operated	(d) Thus, also, the suggestion as to Sexual Selection, or of any other supplementary cause of modification, must be ruled out; and, as in the case of the Lamarckian factors, their co-operation deemed impossible
(e) No detriment arises to the theory of natural selection as a theory of the origin of species by entertaining the possibility, or the probability, of supplementary factors	(e) The possibility, and *a fortiori* the probability, of any supplementary factors cannot be entertained without serious detriment to the theory of natural selection, as a theory of the origin of species
(f) Cross-sterility in species cannot possibly be due to natural selection	(f) Cross-sterility in species is probably due to natural selection

second interpretation would lead to endless reminders that natural selection *was not the exclusive agent*. And, just such an interpretation was upheld both by Darwin's opponents, like St. G. Mivart, and by self-confessed Darwinians like George Romanes. Mivart maintains that

> natural selection acts, and indeed must act, but that still, in order that we may be able to account for the production of known kinds of animals and plants, it requires to be supplemented by the action of some other natural law or laws as yet undiscovered.[25]

During his career, George Romanes established himself as the reasonable opponent of Wallace and Weismann, with the *Times* declaring him "the biological investigator upon whom, in England, the mantle of Mr. Darwin has most conspicuously descended". Coining the terms "neo-Darwinism" and "ultra-Darwinism", with which he labelled Wallace and Weismann's thought, Romanes was extremely attentive to studying the various Darwinian legacies in their full plurality, particularly when it came to the place of natural selection among the causes of organic evolution. His views, insofar as they conflicted with Wallace's, led to diverging definitions of Darwinism, a fact analysed by Fern Elsdon-Baker.[26]

Romanes broke a Darwinian taboo by boldly asking whether natural selection was the only, or even the main cause underpinning the origin of species. For him, Darwin's answer to this question was at once "distinct and unequivocal": "He stoutly resisted the doctrine that natural selection was to be regarded as the only cause of organic evolution". Romanes shored up this answer by asserting the importance of "Lamarckian factors".[27] For Romanes, Darwin's selectionist posterity hardens and deforms the Darwinian legacy, "whether the misrepresentation be due to any unfavorable bias against one side of his teaching, or to sheer carelessness in the reading of his books". In this,

> not only do the Neo-Darwinians strain the teachings of Darwin; they positively reverse those teachings—representing as anti-Darwinian the whole of one side of Darwin's system, and calling those who continue to accept that system in its entirety by the name "Lamarckians".[28]

In particular, Wallace is accused of obscuring certain parts of Darwin's text. Defending the exclusive agency of natural selection is not a Darwinian trait but a Wallacian one. Explaining organic evolution through natural selection alone is a conceivable theoretical position, though there is no justification for attaching Darwin's name to it: the thrust of such a position is instead to "*out-Darwin Darwin*" and would thus be much better served by the label *Wallacism*. It is in this that, for Romanes, the "ultra-Darwinians" deform Darwin[29] and are thus absolutely comparable to another, symmetrical endeavour to out-do Darwin; namely the one attempted by the American neo-Lamarckians. Romanes reads Darwin as an invitation to go beyond natural selection by searching for the other "means of modification" and, by this, opening a programme of critical study. This would conclude with either the rejection or the reassignment of the relationships between natural selection and the origin of species.

To explain the origin of species is to understand how the transformation occurs from distinct but interfertile varieties to distinct and mutually infertile species. From the time of the *Origin*'s first publication, and right up until today, readers have wondered whether Darwin's work actually answers the question raised by its title, taken to be the question of *speciation* (i.e. concerning the mechanisms that prevent species from reproducing with each other).

One of the criteria distinguishing species from variety is the interfertility of individuals from within one species and the sterility of crossings from without. Therefore, the distinction between species and variety must reside in a certain reproductive isolation. If, furthermore, the idea that varieties are *incipient species* be accepted, then a certain porousness between species and variety must be admitted. The real lesson in Darwin would then be that species are nothing more than pronounced varieties, on the one hand, and nascent genera on the other. But how does a species (which, in principle, is only a variety) find itself reproductively isolated from others? Such an interpretation hints at re-framing the *origin of species* question as the *speciation* question. On the basis of this understanding of the origin of species, readers often state that the book does not answer its own question.

If species really are only "marked" varieties (separated by a barrier of sterility), then explaining the origin of species (taken to mean *speciation*) just is a case of explaining how this barrier of sterility came about. Within the Darwinian framework, this equates to establishing whether natural selection and, more generally, descent with modification can actually produce this inter-species barrier. Yet Darwin says almost nothing of this. This difficulty has been called "Mr. Romanes' paradox" (see below).[30]

Origin of species by migration (Wagner)

Darwin devoted two chapters of the *Origin* to geographical distribution: Chapters XI and XII of the first edition. Geographic distribution constituted one possible explanation for reproductive separation between species: with its focus on geographic barriers, limiting the range of each life form, distribution gives rise to the isolation of individuals in different environments. For several commentators, it was Darwin's focus on the issues of geographic distribution that may have been "the original stimulus for Darwin's conversion to evolution", as Peter Bowler has put it.[31] He constantly set the *vera causa* of ordinary generation combined with migration from one single location in opposition to the miracle of special centres of creation.[32] He analysed the relationship between island species and neighbouring mainland productions: for instance, on the Galapagos Islands, "almost every product of the land and water bears the unmistakeable stamp of the American continent".[33] Darwin concluded that the island forms had been established by occasional migration, followed by divergence, in accordance with his theory of *descent with modification*.

This emphasis on the role of geographic isolation was certainly the explanation for species formation advanced by what D. Depew and B. Weber call, a "dissident Darwinian tradition", going back to Moritz Wagner and George Romanes, "which stressed the role of geographic isolation of populations in speciation".[34] In this tradition, the focus is no longer on the mere *transformation of species*, but also on the *diversification of species*, as Ernst Mayr put it.[35] If Darwin was aware of this difference between transformation and diversification as two components of evolution, a clear distinction was only achieved with the work of two post-Darwinians, J.T. Gulick and George Romanes.[36]

In the post-Darwinian literature, the Munich professor Moritz Wagner (1813–1887) was both praised for putting forward the idea of geographic isolation as an important factor in speciation and also blamed for combining it with his own peculiar conception of variations. In particular, Wagner thought that migration and subsequent isolation would trigger increased variability among individuals.[37] A passionate reader of the *Origin of Species*, Wagner concluded that Darwin's *natural selection* belonged to the programme nineteenth century naturalists had set themselves: the search for simple laws. Wagner himself thought he was simply extending this programme when he proposed not a simple *phenomenon* of migration, but rather a *law* of migration constituting the necessary condition for natural selection.[38] Observing the laws that assure the distribution of organisms, Wagner noticed "mysterious phenomena" whose causes remained unknown to him, particularly in cases concerning the habitat of several animal species. When a barrier as flimsy as a narrow river can separate two varieties of beetle, for example, it is difficult to imagine any Providence who would have assigned each variety its own habitat through specific acts. Likewise, when tall mountains act as borders between species, the climatic causes cannot explain those facts which remain, on the whole, out of the reach of Darwinism's three fundamental ideas: individual mutability; transmission of novel characteristics through descent; and conservation and reinforcement of these characteristics in a certain direction over generations—all effected through the *struggle for existence*. This is why Wagner, without turning to the action of exterior conditions, and without relating it to the competition between organisms, aimed to complete Darwinism with a new natural law. His contribution to the theory of the origin of species took the shape of a "law of the migration (*Migrationsgesetz*) of organisms", exceptional for its simplicity: "Like all natural laws or causes of phenomena, this law is remarkable for its simplicity, for it is based on the two most powerful impulses [*Trieben*] of all living beings, *viz.* self-preservation and reproduction".[39]

Because of competition, individuals must constantly go beyond their distribution zone. They manage this through either voluntary displacement or through passive migration (displacement by water or wind, for instance). Wagner identifies a "tendency to emigration (*Tendenz der Wanderung*, *Migrationsstreben*)" attached to the "efforts (*Streben*) of all organisms to conserve and multiply themselves". Random chance can also bring about numerous migrations and indirectly eliminate free crossings. Wagner illustrates this using the scorpions that were found in the wooden casing of the obelisk (now installed at Place de la Concorde) when it arrived in Paris in October 1836.[40]

Wagner's intention is to show how natural selection and the law of migration are very closely tied: migration allows natural selection to originate species. Without the isolation produced by migration, variations would be lost and species would not originate. This connection between law of migration and natural selection provides a response to Cuvier's anti-transformist objection that the ibis and the crocodile have not changed for 4,000 years. For Wagner, this is clearly related to the fact that they did not migrate, that they never changed habitat: without migration, no isolated colony can be formed and no selection can occur. However, comparing the Nile crocodiles with those of the Ganges, or even with alligators, we easily grasp that in circumstances of geographical isolation differences can be fixed into form.

Furthermore, Darwin himself had acknowledged that many objections to his theory fell apart thanks to the law of migration. The point of contention with Wagner involves isolation and its modalities: is it merely useful? Or, as Wagner suggests, is it genuinely necessary? For Darwin, species originate on the wide continental expanses.[41] For Wagner, origination requires migration, not strictly speaking an "isolation" (in the sense where a population finds itself trapped on an island, *isola*), but rather a separation or segregation: with populations migrating, variations find themselves de facto separated, with no need for natural selection to play its role of eliminating and sifting.

Origin of species by "physiological selection" (Romanes)

By proposing a "law of migration", Wagner was asking whether natural selection did or did not explain the origin of species. The problem here is not the *existence* of natural selection (does nature select?) nor its *adequacy* (what does it act upon?), rather it is a question of its *responsibility*: what does it enable us to explain?[42] Most especially, once this process has been elucidated, does it account for the origin of species?

George Romanes is to be counted among those authors who, though they may not reject the existence of natural selection outright, nevertheless endeavour to dissociate it from the origin of species. Romanes assigns it another function. Certain of Darwin's successors did not take selectionism on board; they relegated natural selection, made it secondary, and also reassigned its functions— natural selection originates not species but rather some other thing. For Romanes, for instance, this other thing is *adaptations*.[43] Again, Romanes stakes his claims on Darwin's own authority:

> Mr Darwin himself has freely acknowledged that his theory of natural selection is not in itself a sufficient explanation of the origin of species. He therefore supplemented the natural causes which are together comprised under this term by sundry other causes of similarly natural kind.[44]

Romanes maintains that the theory of the *origin of species by means of natural selection* falls down at three major hurdles:

1   The sterility of inter-species breedings: why does the origin of species lead to the mutual sterility of individuals belonging to different species whereas simple variation, generally speaking, remains compatible with inter-fertility?
2   The origin of variations: why does the same useful variation emerge simultaneously in a large number of individuals?
3   The characteristics defining species; they are so tiny that their utility must be questioned.

These three objections question the relation between species and selection by differentiating two levels, one morphological (anatomical difference) and one physiological (reproductive barrier). For Romanes, the difference between species is of the same *morphological* magnitude as the difference between two varieties, yet the first forms an insuperable *physiological* barrier between two forms that may be anatomically similar to each other. Furthermore, the difference between species is neither the simple difference between two varieties nor the change of body plan that characterises differences of genus or class.

This method employs a two-pronged approach to species, in terms of resemblance and in terms of inter-fertility. It is further combined with comparative anatomy and physiology. From a morphological point of view, the origin of species signifies a process of modification in organic types, and there is nothing which *necessarily* demands that variation in the type imply different degrees of mutual sterility. Moreover, natural selection works on *useful* variations. But, according to Romanes, one species differs from another by a set of altogether minor marks, when compared to those marking the differences between genera, families, orders, and classes.[45] If natural selection *originates* species, then these imperceptible differences *must* include some utility which does not escape Nature's vigilance. For Romanes, this leads to simply supposing such "utilities" (which remain unknown to us) in order that the differences be susceptible to natural selection. Moreover, species originate by "divergence" or "diversification of character", what Romanes calls "the ramification of species".[46] For Darwin, the more diversified the descendants, the more they will occupy distinct places in the economy of nature and therefore be more likely to multiply.[47] Can natural selection account for the secondary importance and utility of characteristics leading to speciation as well as the ramified disposition of the species? For Romanes, the marks that differentiate species are not useful variations and therefore escape natural selection. Consequently, natural selection cannot originate species More strictly speaking, natural selection is, in fact, a theory of the origin of *adaptations*, whatever their nature (morphological, physiological, psychological) or taxonomical level (genus, family, order, class ...) may be.

Romanes claims not to attack natural selection but instead says that he defines its characteristics and functions and, thereby, dismisses "misnomers". If natural selection does explain the origin of adaptations, and not the origin of species, then other factors must be found to explain the latter. In Darwin's case these are use and disuse, correlative variation, and sexual selection; in Wagner's case, it is the law of migration that must be included. Romanes presents "an additional factor" to explain the formation of specific types, something he names "the Prevention of Intercrossing with Parent Forms, or the Evolution of Species by Independent Variation", or again "physiological selection, or segregation of the fit", whose role it is to complete "natural selection, or survival of the fittest".[48]

Romanes' position on natural selection and its influence is exactly symmetrical to that held by, for example, the botanist Carl Nägeli.[49] Nägeli implicitly shifts the question from the origin of species (*Entstehung der Arten*) to the essence and origin of living substance (*Wesen und Entstehung organisirten lebenden Substanz*); then, having highlighted the causes of phylogenetic change, he studies not the origin (*Entstehung*) of species, but still only these changes (*Veränderungen*) to organised substance. In addition, he deviates explicitly from explanation by means of natural selection, which takes the formation of species to be a consequence of the formation of races: on the contrary, Nägeli thinks the two levels of formation (races vs species and varieties) must be distinguished. He believes that natural selection can explain the origin of species but not the origin of body plans; Romanes attributes to it responsibility for body plans but not species.

In light of all of the above, the nature of "Mr. Romanes' paradox" should be clear. Romanes is a major figure of Darwinism, despite denying natural selection's role as the means behind the origin of species. How can this be possible? Giving the Romanes paradox its due consideration, must we then admit the existence of "Darwinians" who refuse to see natural selection as the origin of species?

Romanes' conference in May 1886 provoked numerous reactions. Many came away understanding his message to mean that natural selection originated nothing, especially not species.[50] Others saw it as a healthy purging of Darwinism.[51] Romanes made it clear that he did not wish to question natural selection's pertinence, but rather to simply localise its utility. The originality of his position is in the distinction he makes between two domains: origin (or "genesis") of adaptations and origin of species. The mechanical theory of natural selection does not originate species. Although "it has hitherto been entangled on account of its having been made to 'pose' as such", it is in fact "a theory of the genesis of adaptive structures and instincts".[52] But can one, having so narrowly localised the action of natural selection, still claim to be "Darwinian"? Romanes thinks it absolutely consistent and even reiterates the point: "natural selection ought not in strictness to be regarded as a theory of the origin of species, but rather as a theory of the development of adaptive modifications".[53]

Natural and physiological selections are reconciled because the difference between species does not consist exclusively in useful characteristics and, therefore, natural selection does not act on this level. Darwin's son, Francis, played an ambiguous role in this debate. On the one hand, he fed Romanes' Darwinian fire by declaring that the first expressions of physiological selection were to be found in a sentence added to the fourth edition of the *Origin*; but on the other hand, he also put distance between his father and physiological selection.[54] Ultimately, his intention was less to show that his father would have agreed with Romanes but more so to deny Romanes any claim to being the first to make the discovery.[55]

On several occasions, Romanes attempted to reconcile the text Francis Darwin had quoted from the *Origin* with his own theory of physiological selection.[56] To this end, he shows that the incriminating passage from the *Origin* is located within an argument showing that sterility between species cannot have been produced by natural selection. Even if it be supposed that sterility between descendants and the parental forms constituted an advantage, natural selection would have no hold over it and sterility would have to be explained by some other means. Above all, Romanes emphasises the difference between Darwin and himself: Darwin considered sterility to be either an advantage or a drawback in varieties whose differentiation had already taken place elsewhere (or else was currently under way) and was to be preserved by the non-occurrence of mixes; Romanes, on the other hand, considered the way in which a rupture in intergenerational fertility can, in and of itself, constitute the origin of species.

Wallace, on the other hand, contested Romanes' thesis with a certain number of facts and called for a distinction to be made between the fertility or sterility of *first crossings* and those of *hybrids*.[57] For instance, the crossing of a donkey and horse is fertile (producing the mule), but the hybrid produced by this crossing is, as is the case with many hybrids, infertile. Wallace's objections also concern "the assumption ... that the same variation occurs simultaneously in a number of individuals": Wallace adduces "evidence—copious evidence—that the supposed assumption represents a fact, which is now one of the best-established facts of natural history".[58]

What then, ultimately, is the Darwinian theory? On this point, the botanist W.T. Thiselton-Dyer railed against Romanes and his ambiguities: it is too often ignored that this theory is found in a work whose title is not just "*On the Origin of species*", but "*On the Origin of species by means of natural selection*". This is a clear statement of intent from the book's author. Yet, had Romanes not just asserted that natural selection originated not species but adaptations? Thiselton-Dyer's protest to this consisted in saying that, by tearing the relation between natural selection and the origin of species out of Darwinism, Romanes removes its very substance.[59] The Romanes paradox then is this: how can "*the biological investigator upon whom, in England, the mantle of Mr. Darwin has most conspicuously descended*" simultaneously deny that the Darwinian theory has reached its goal? The very goal—to have proved that natural selection really is the means behind the origin of species—that its title had so boldly announced?

Another bone of contention concerns the adaptive (or non-adaptive) character of specific differences. Romanes defends physiological selection through various arguments, notably the "argument from the inutility of specific differences": "why is it that apparently useless structures occur in such profusion among species, in much less profusion among genera, and scarcely at all among families, orders, and classes?"[60] If natural selection explains the origin of species, then this implies that specific differences are adaptive. Yet, in the majority of cases, they are not; this reduces the action of natural selection to only those (rare) cases where these specific differences are adaptive. Everything hangs on affirming whether differences in species are adaptive or not, or in other words, whether they are useful to any kind of degree or are otherwise utterly useless.

This is the line that separates Romanes from his opponents. For Wallace, "Romanes makes a great deal of the alleged 'inutility of specific characters'", but "there is no proof worthy of the name that specific characters are frequently useless";[61] for Thiselton-Dyer, if the inutility of *specific features* is established, then Romanes has "inflict[ed] a deadly blow on the Darwinian theory, the very essence of which is that specific differences must be advantageous".[62] Darwin drew the "conclusion that all specific differences in plants are probably adaptive". Of this, Thiselton-Dyer thinks "it seems only a reasonable induction, the validity of which is strengthened every day by fresh observation".[63]

It turns out, then, that the difference between Romanes and Darwin is not just in the detail but constitutes, in fact, a divergence of method, almost of scientific mindset: Darwin proceeded by naturalist method, through the patient accumulation of facts, but Romanes, "on the other hand, frames a theory which looks pretty enough on paper, and then, but not till then, looks about for facts to support it".[64] Romanes advanced no proof for variations leading to sterility with the parental form yet maintaining fertility within the variety, which would indeed be a highly improbable case. In the end, even Romanes' terminology is fallacious, since what it refers to as "physiological selection" would more fittingly be called "reproductive isolation"; as for the expression "segregation of the fit", it seems quite unsuitable for referring to the variations "of an unuseful kind" that are Romanes' focus. As for the "indifferent" variations; "correlated variation does give rise to a large classification of non-significant characters".[65]

Such debates were not aimed at determining who was for or against Darwin; rather, they sought to establish who was genuinely Darwinian. What is it to be faithful to Darwin? In arriving at a response, various versions battle it out. Romanes pokes fun at the multiple accusations: "unsustained generalizations" that he has "shrivel[ed] up the Darwinian theory to very small dimensions" alongside claims that he has "roundly denied it altogether".[66]

When Thiselton-Dyer accuses him of drawing out a "strained" interpretation of Darwin's writings, Romanes throws the accusation back by demanding to know: who is purifying Darwin?

> Over and over again—and more and more emphatically the later the editions of his works—Mr. Darwin insists that he does not regard natural selection as the only agent which has been concerned in the origination of species, and therefore concludes—to quote only one additional passage from among many to the same effect:

> *No doubt the definite action of changed conditions, and the various causes of modification, lately specified, have all produced an effect, probably a great effect, independently of any advantage thus gained.*[67]

Romanes recalls Darwin's complaint of "steady misrepresentation"[68] on this very point. And yet it is still this very doctrine, affirming that natural selection alone rules all, a doctrine the man himself constantly rejected, that posterity has chosen to call "pure Darwinism". Even though Darwin refused the doctrine of pan-utilitarianism in a section of Chapter VI ("Utilitarian doctrine how far true"), Wallace describes it as "a necessary deduction from the theory of natural selection".[69] It is this same pan-utilitarian doctrine again that Weismann's school prioritised and which provided the foundation for Thiselton-Dyer's objections against physiological selection.

The concept of physiological selection claims to fill the gaps in Darwin's system, necessarily bringing about a reconfiguration of "Darwinism" itself: it exposes how species originate, through sterility between the species and the variety that follows from this; precisely, this is what "M. Romanes' theory provides and what Darwin's theory did not give".[70] Natural selection exists and its role is important; it is what builds wings. But it is not down to natural selection to differentiate two species of swallow. From Romanes' perspective, natural selection is a theory of adaptations. Rather than being a theory of species, it underpins the origin of genera, families, orders: physiological selection, on the other hand, thanks to its focus on the inter-species reproductive barrier, could lay claim to the title of "theory of the origin of species".[71]

Natural selection and auxiliary principles (Hartmann)

How far does the causal efficiency of natural selection extend? And, correlatively, how much does Darwin's theory account for? Darwin's explanatory system must face two objections. On the one hand, natural selection uses variations to explain transformations in organic forms, but it does not account for the determinations of these variations; it shapes a material which it seems tasked to conserve, but it is a material it does not seem capable of producing. This is why, according to the philosopher Eduard von Hartmann, natural selection constitutes the heart of Darwin's system but is only an "auxiliary hypothesis of transformism", while the veritable constitutive element of transformism is variability or the origin of variations.[72] On the other hand, natural selection explains adaptations or physiologically useful characters but does not account for a set of purely morphological characteristics, whether these be devoid of utility or even outright harmful. Because of this, Darwin's theory, threatened with the admission of its incompleteness, must turn to certain "auxiliary principles" to shore up natural selection.

According to Hartmann, these principles are four in number. The first two—influence of exterior circumstances and effects of use and disuse—concern the causes of variation (the material natural selection shapes); the next two—sexual selection and correlation of variations—concern the explanation of useless or harmful characteristics (left unexplained by natural selection). These four principles function as a body of secondary hypotheses developed by Darwin in the *Origin*: some (e.g. sexual selection) *corroborate* natural selection, others (e.g. correlation of growth) *complement* it. Their purpose is to explain the facts of variation and of inadaptation. In principle, they act upon a different domain to selection and don't at all affect the value of this concept: their purpose is only to complete or confirm it. Nevertheless, the paradox of the "auxiliary principles" is that, arriving as a reinforcement to natural selection, they actually end up draining it of its content.

The theory of natural selection, the pillar of the system, seems undone from the inside, and doubly so: first, by the fact that variability is not the result of chance but answers to laws; and, second, by the fact that the forms of living organisms are not fully legible through the lenses of only adaptation and utility, leaving room for inutility, if not to say, potentially at least, harm. This constitutes a criticism of the Darwinian *agent* (natural selection *disposes* of forms) through study, in parallel, of the *material of agency* (variation *proposes* forms) and the *product of agency* (the structures which are *set down*). In this framework, the intervention of other factors, far from propping up the Darwinian system, would result in a weakening of the very core of natural selection and the theory of a coherent, gradual transformism along with it.

Ultimately, these principles lead us back to the laws of variation and create counter-productive knots within the Darwinian system. Such fall-back positions, intended to protect the project's core, end up undermining its legitimacy by revealing that natural selection has no agency, or that its agency depends on other principles. This logical, or chronological, priority status opens the way for the auxiliary principles to be thought of as more fundamental—and, therefore, to become themselves primary.

The auxiliary principles question the utility of organs. What is useful serves a purpose; what is useless falls into disuse. The influence of environment and exterior circumstances raises the question of an organism's adaptation to its environment. In such a context, utility intervenes as that variation which enables survival, through the organism's physiological adaptation to exterior conditions. By supposing an action of exterior circumstances, we admit that the environment is possessed of a power to sanction useful variation, perhaps even to produce it; in other words, a power to influence variation in the direction of utility. It is, therefore, a manner of affirming that variation is not random but directed towards what is useful. And, in any case, are all organs really useful? Confronted with this uncertainty, and in the eventuality that useless or harmful characters may finally have to be accounted for, Darwin called on two other auxiliary principles: sexual selection (why peacocks are decorated with ocelli) and the correlation of variations (why albino cats are also blind).

Within this dimension of auxiliary principles, we find an important element in support of our general hypothesis of a Darwin that can always be played against himself. The auxiliary principles do not conflict with Darwin; they are taken from within the *Origin* and Darwin's other works themselves. They are employed a posteriori to undo Darwinism from the inside, draining it of its substance. In the name of the laws of variation that Darwin decided to seek, they suggest the impossibility of random variability and, through this, they empty natural selection of both the variations that drove it and its effective power to shape the lines of evolution. The auxiliary principles weaken natural selection even as they profess to flesh it out.

Hartmann proposed his own schema, assigning each of the four auxiliary principles a place within the mechanism of gradual transformism that Darwin offered in the *Origin* (Table 6.2): the principle of natural selection is divided into its three components; the four auxiliary principles are outside of the mechanism of natural selection, but they seem to be contradicting the components of natural selection from within the body of Darwin's work. For instance, correlation of growth seems to conflict with variability, use and disuse with hereditary transmission, etc.

The process of natural selection, *sensu lato*, takes up a strictly mechanical first level (selection via the struggle for existence) to which two other law governed levels are added: hereditary transmission and variability. Hartmann brings to light how, under the umbrella term "natural selection", are in fact regrouped not only the struggle for existence but also many other as of yet poorly distinguished elements. The Darwinians believe these can just be left in the shadows, but Darwin's critics drag them centre stage and lay them out for all to see; they suppose that the Darwinian system will crumble just as soon as a single one of these badly disentangled elements—chance variation for example—no longer stands.

For Hartmann, Chapter V of the *Origin* indicates clearly enough that variability is not random, but "conforms to a plan". Therefore, with inheritance and variability we are already dealing with manifestations "of the internal law of evolution". Still, according to Hartmann, the four auxiliary principles allow Darwin to complete the general schema of "gradual transformism". The schema of "kinship brought about through descent" is not limited to this gradual transformism but also includes "heterogeneous generation", where transformism advances by leaps or by metamorphosis, in conformity with a law of development.

In this way, the theory of transmutation by natural selection (understood as a gradual mechanism) is reinterpreted as one form of the theory of descent, of which the theories of development make up the rest. In its Darwinian form, the theory of descent postulates that organic forms are related to each other through bonds of filiation; but among the theories of descent, beyond natural selection alone, we must also count the *Entwicklung* theories, i.e. theories of heterogeneous generation which advance that forms metamorphose into each other without natural selection playing any role. Finally, the "ideal kinship" between types must be accounted for in order to arrive at the full schema of the "theory of organic evolution". In reality, Hartmann wanted to show how the auxiliary principles, as well as the importance of heredity and variability, guided Darwin from a strictly mechanical procedure (selection's sieve) to "an internal law of

Table 6.2 Natural selection among the other evolutionary mechanisms

Systematic parenthood of types								
Real parenthood, effected by descent								Ideal parenthood realised by analogies in regular evolution
Gradual transformism*							Heterogeneous generation, by metamorphosis of the germ, according to a law	
Natural selection			External circumstances directly acting on a pre-existing inner tendency to modification**	Role of use and disuse, according to instincts or conscious teleological activity**	Sexual selection by instinct of sexual preference (unconscious ideas of types)**	Law of correlation of growth and modifications**		
Selection in the struggle for existence	Hereditary transmission of individually acquired characters	Variability, conforms to a plan in its direction, intensity and correlation						

Various manifestations of the internal law of evolution

Theory of organic evolution

Mechanical process

Source: adapted from Hartmann 1877.

Notes

* "Gradual transformism" is how Hartmann described Darwin's system.

** These are the four auxiliary principles that Darwin introduces in his works in order to complement the action of natural selection.

evolution or development", suggested in particular by the existence of organs which do not follow the principle of utility as well as by the refutation of randomness entailed by affirming laws of correlation.

Hartmann didn't reject the theory of gradual transformism: he limited himself to completing it, appealing for evolution to be considered as a product of other mechanisms, notably metamorphoses ("heterogeneous generation"). The fact that organic variations are directly dependent upon changes occurring in the environment does not necessarily imply that variation is *guided* or that it is necessarily adaptive. However, if exterior circumstances do have an influence, then this produces an "internal tendency in organisms towards modification", through which the "internal law of evolution" manifests

> the modification of organisms by exterior influences always supposes a pre-existing aptitude and an internal tendency towards modification, without which the organism would perish or live miserably in a hostile environment, instead of physiologically adjusting itself to the modified exterior environment.[73]

Likewise, the principle of use and disuse seems to relegate natural selection to a subordinate role, instead constituting an argument that is more suited to internal evolution. Indeed, what governs use and disuse is habit: for use to be modified, a modification of the instinct is first required, this would then lead to a modification of use and, ultimately, the potential modification of the organ. For Hartmann, this boils down to introducing an unconsciousness, if not even a teleological characteristic, into the natural mechanics of Darwinism, and also to advancing the idea of an internal law of evolution, primal with respect to any organic or mechanical evolution brought about after the fact by use and disuse. Through the agency of a system of needs, this would seem to place a mind-first principle at the base of evolution. Finally, as Hartmann remarks, whatever the cause may be—law of organic evolution, instinctive necessity, conscious intellectual activity with an end in mind—the principle of use and disuse appears to be a "technical expedient for the accomplishment and acceleration of steady internal evolution". If evolution (phylogeny) is concerned with morphological changes, then adaptation proceeds by physiological adjustments within a single given morpheme. Therefore, adaptation is not the *primum movens* that triggers evolution; for Hartmann, that principle is to be sought in an "internal impulsion". This is where the "insufficiency of the utilitarian principle" in accounting for the phylogenetic tree stems from: "Here we see ourselves summoned back to an internal law of evolution; that the gap between one type and another must be filled either through heterogeneous generation or through gradual and steady transformism".[74] Hartmann's schema suggests that Darwin had pinned the fate of his system on supplementary hypotheses which may seem superfluous. I will develop two examples: gradualism and sexual selection.

In a Darwinian framework, natural selection sees its fate locked to *gradual* transformism: it is combined with *continualism* and the *absence of leaps*. And yet, the "descent with modification" context is broader than this, including also heterogeneous generation. This is why certain critics, like the German botanist Albert Wigand, who considered the gradualist dimension of Darwinism to be untenable, went ahead and threw natural selection out with it; this is also why some disciples, like T.H. Huxley, explicitly set aside gradualism and admitted evolutionary leaps. In reality, gradualism is not the only possible avenue for the theory of selection and, once this has been admitted, the theory of natural selection seems capable of following its own fortune, independently of the initial context given to it in Darwin's words. For Hartmann, "natural selection applies equally well to types resulting from heterogeneous generation as it does to those whose source would be a series of imperceptible and chance modifications".[75] Criticising Darwinian gradualism does not necessarily imply casting natural selection aside, but it does entail making it autonomous; liberating it for use in new applications. In recent times, Stephen Jay Gould's attitude stands as a good illustration of this point.

Sexual selection is also considered a necessary part of natural selection by orthodox Darwinians. Weismann, for one, supports this view of sexual selection as a corroborating element for natural selection: "sexual selection goes hand in hand with natural selection".[76] However, others like Argyll or Hartmann consider that sexual selection weakens natural selection. A similar argument was developed by Vernon Kellogg in 1907. He presents sexual selection as "one of Darwin's supporting theories":

> it is based on a postulated particular and limited kind of natural selection, not involving determination between life and death, but a determination between going childless and leaving posterity,—which is, after all, the essential determination in general natural selection.[77]

But far from being true to the spirit of natural selection, Kellogg suggests that the theory of sexual selection "has a much less mechanical and automatically working basis", even involving assumptions about the aesthetic sensibilities of animals. For Kellogg, sexual selection is "one of the first Darwinian outworks to be sadly breached by attack".[78]

It is notable that many of Darwin's opponents (including Hartmann) did not radically reject natural selection, some even granted it a place within their theories. They just didn't accord it the same "origin" role that Darwin had. The same judgement can even be extended to include certain of Darwin's defenders (Romanes for instance), so that the idea of an opposition between "Darwinians" and "anti-Darwinians" loses much of its clarity and force. Instead of an "eclipse of Darwinism", what we have is the blossoming of diverse Darwinisms, mutual rivals, each playing Darwin against Darwin in their own particular way.

For many readers of the *Origin*, species originate through the establishment of a physiological, interfertility barrier—a phenomenon that twentieth-century biologists called "speciation". However, Darwin leaves this outside of his analysis. Can it then be said that "natural selection" explains the origin of species?

or does it explain something else, such as the origin of adaptations and large body plans? It falls to others to propose principles that will account for the inter-species barrier: "law of migration" (Wagner), "physiological selection" (Romanes), and so on. Do these complementary principles reinforce Darwin's stance? Or do they drain it of its substance? If natural selection cannot be said to be the actual "origin of species", for as long as it does not account for interspe-cific barriers, then it accounts only for the "origin of adaptations". Was this really what Darwin had in sight?

A similar question arises from within the *Origin* itself. Besides natural selec-tion, Darwin develops several auxiliary principles whose supposed purpose is to reinforce the defensive walls of his core idea: these are the principles of sexual selection, correlation of growth, etc. By admitting these principles in order to account for apparently useless characteristics, Darwin opens the door to other modes of evolution. No longer proceeding through the gradual accumulation of differences produced over the course of sexual generation but through laws of internal development, evolution might also be effected through simple metamor-phosis, or as a psychological force, perhaps even a metaphysical principle.

Notes

1 The treatment of Chapter V in the two *Cambridge Companions* devoted to Darwin and his masterpiece is quite revealing here. Richards and Ruse's *Companion* (2009) strangely breaks with its step by step reading of the *Origin* by suddenly jumping from David Kohn's focus on divergence (commenting chapter IV) to A.J. Lustig's dif-ficulties (dealing with Chapters VI and VII), or with Robert Olby's paper on variation, which only cursorily evokes chapter V. Hodge and Radick's *Companion* (2009) offers a better treatment with Kenneth Waters' presentation of the *Origin's* various "argu-ments", dealing with Chapter V as a challenge to the idea of "one long argument". However, Waters remains unclear as to what to actually do with Chapter V itself: see especially Waters 2009, table on p. 125.
2 A good example of complete rejection of Chapter V can be found in Bates and Hum-phrey (1957, p. 165), which provides readers with excerpts of chapters of the *Origin*: "Chapter V of the *Origin*, dealing with the 'Laws of Variation', has been omitted entirely because the development of the science of genetics puts this subject in a quite different perspective from the one it had in Darwin's day". A case of embarrassment caused by Chapter V can be documented in Reznick 2010 (pp. 102–103): "Because of Darwin's lack of understanding of inheritance, however, many portions of this chapter are archaic. For the sake of brevity, I will present synopses of only some parts of the chapter".
3 *Origin* 1859, p. 486, *Var* 255 (#235).
4 *Origin* 1859, p. 150, *Var* 298 (#147).
5 *Origin* 1859, p. 134, *Var* 278–279 (#27–33).
6 *Origin* 1859, p. 159, 352 and 482.
7 *Origin* 1859, p. 165, *Var*.316 (#269).
8 *Origin* 1859, p. 154, *Var* 303 (#186).
9 *Origin* 1859, pp. 131–134, *Var* 275–279.
10 *Origin* 1859, p. 172, *Var* 322 (#15).
11 See above, Chapter 2.
12 Darwin to Bronn, 14 February 1860, CCD 8 83.
13 *Origin* 1859, p. 6, *Var* 75 (#50): "*the main*" is changed to "*the most important*" in the fifth edition.

14 *Origin* 1872, p. 421, *Var* 747–748 (#183).
15 *Origin* 1872, p. 421.
16 Darwin 1871, vol. 1, pp. 152–153. He returns to this in the preface to Darwin 1874, p. v.
17 Moore 1979, p. 127.
18 Young 1985, p. 119.
19 Gayon 1998, pp. 408–409.
20 For instance, Bastian 1872, vol. 2, p. 578.
21 Candolle 1873, pp. 10–11.
22 Darwin 1868, vol. 1, p. 10.
23 Wallace 1889, p. 444 (emphasis added). See also pp. vii–viii.
24 *Ibid.*, p. 148.
25 Mivart 1871, p. 5.
26 Elsdon-Baker 2008.
27 Romanes 1895, pp. 2–3.
28 *Ibid.*, pp. 8–9.
29 Cf. Table 6.1 for a comparison of Darwinism and Wallacism.
30 Thiselton-Dyer 1888.
31 Bowler 1989a, p. 192. See also Limoges 1970.
32 Bowler 2009, pp. 153–172.
33 *Origin* 1859, p. 398.
34 Depew and Weber 1995, p. 278.
35 Mayr 1982, p. 400.
36 See Mayr 1982, p. 400, where he credits Gulick 1888 for the distinction between *monotypic* and *polytypic* evolution; and Romanes 1892–1897 for a distinction between transformation in time and transformation in space (diversification).
37 Sulloway 1979; Mayr 1982, pp. 562–563.
38 Wagner 1873, p. 5.
39 *Ibid.*, p. 3.
40 *Ibid.*, p. 34.
41 This opposition between Darwin and Wagner was emphasised by Mayr 1982, p. 563.
42 Hodge 1977, pp. 237–246.
43 Romanes 1886, p. 360.
44 *Ibid.*, p. 360.
45 *Ibid.*, p. 362.
46 *Ibid.*, pp. 362 and 363.
47 *Origin* 1859, pp. 111–116, *Var* 205–210 (Romanes 1886, p. 363, quotes the sentence #263 but replaces "polity of nature" by "economy of nature").
48 Romanes 1886, pp. 315–316. See also Romanes 1888, p. 174.
49 Nägeli 1884.
50 Argyll 1886, pp. 335–336.
51 For example, *The Philosophical Review*, vol. 5, no. 6 (November 1896), p. 667.
52 Romanes 1886, p. 360.
53 *Ibid.*, p. 408.
54 *Origin* 1866, p. 311, *Var* 444 (#159.6:d).
55 Cf. *Nature*, 34 (2 September 1886), p. 407.
56 *Nature*, 34 (9 September 1886), p. 43 (7 October 1886), p. 545.
57 Wallace 1889, pp. 180–184.
58 Wallace, "Physiological Selection and the Origin of Species", in *Nature*, 34, (16 September 1886), p. 467.
59 Thiselton-Dyer 1888, p. 8.
60 Romanes 1886, p. 362.
61 Wallace, in *Nature*, 34, (16 September 1886), p. 467.
62 Thiselton-Dyer 1888, p. 8.

63 *Ibid.*, p. 8.
64 *Ibid.*
65 *Ibid.*, p. 9.
66 Romanes 1888, p. 173.
67 *Origin* 1872, p. 160, *Var* 368 (#210–212:f), quoted in Romanes 1888, p. 173.
68 *Origin* 1872, p. 421, *Var* 748 (#183.0.0.4:f).
69 Wallace 1870, p. 47.
70 *Revue scientifique*, XIII (9 April 1887).
71 For a recent presentation (and defence) of Romanes' physiological selection, see Forsdyke 2001, pp. 47–63.
72 Hartmann 1877, pp. 67 and 112.
73 *Ibid.*, p. 114.
74 *Ibid.*, p. 88.
75 *Ibid.*, p. 68.
76 Weismann 1910, p. 42.
77 Kellogg 1907, pp. 16–17.
78 *Ibid.*, p. 17. See also the whole of Chapter 5 in Kellogg.

Part III

Radical origins

Darwin-the-Cosmologist

In this final part, I examine how the *Origin* was read not as "one long (epistemological) argument", but as a metaphysical project.

Many readers claimed that Darwin had provided the foundations for a new cosmology but complained that he had failed to fully develop it. Enquiring into origins—as highlighted by Darwin in the title of his book—was, in their eyes, a question that called for a methodological break with traditional empiricism.

From the *Origin*, these readers retained only a few words: rather than focusing on the origin of species and its various mechanisms, they enquired into the origin of life and the origin of mankind. These key issues, only alluded to in the last chapter of the *Origin*, demand meticulous analysis: Darwin himself raised the problem, but his *Origin* failed to provide it with the thorough treatment required.

7 "Mystery of mysteries"

The temptation of origin

In the opening passages of his *Origin*, and with grandiose insistence, Darwin proclaims the great significance of the question of the "origin of species" by declaring it to be the "mystery of mysteries". The label was borrowed from "one of our greatest philosophers", whom Darwin nevertheless did not bother to name.[1] We now know that this reference to the "mystery of mysteries" is a gracious tip of the hat to John Herschel.[2] Before the *Origin*, Darwin had used the phrase in previously published works, like his *Journal of Researches*.

However, those various layers of intertextuality were lost on many nineteenth century readers who knew nothing, or almost nothing, of Darwin's intense notebook-scribbling. For instance, Clémence Royer, the first translator of the *Origin* in French, asserted that the turn of phrase was taken from von Humboldt's *Cosmos*—a claim later taken up by Camille Flammarion.[3]

With the hindsight of today's erudite Darwin scholarship, it is easy for us to see that these French readers were unmistakably wrong. But it is nevertheless amusing that none other than von Humboldt should stand in for Herschel in their "ignorant" eyes since, in his *Autobiography*, Darwin states that "Humboldt's *Personal Narrative* … and Sir J. Herschel's *Introduction to the Study of Natural Philosophy* stirred up in me a burning zeal to add even the most humble contribution to the noble structure of Natural Science".[4]

The "origin of species" question hung as a spectre over the naturalist literature of the time, galling the common desire for all remaining mysteries to be laid to rest. J.D. Hooker, for example, mentions it (albeit only to promptly push it aside) in the introduction to his *Flora of New Zealand*. In Darwin's wake, Gustav Jaeger tried to ascertain how the greatest thinkers, since the time of Plato, had approached the question of "this supreme mystery of creation, the origin of animal and vegetable forms". Asa Gray asked "for the reasons which call for this new theory of transmutation" since "[the] beginning of things must needs lie in obscurity, beyond the bounds of proof, though within those of conjecture or of analogical inference". "Why", he asks,

> this continual striving after "the unattained and dim?" Why these anxious endeavours, especially of late years, by naturalists and philosophers of various schools and different tendencies, to penetrate what one of them calls "that mystery of mysteries", the origin of species?[5]

It is this association between origin and mystery that looms over the opening passages of the *Origin*: to enter is to pass through its shadow. Even Owen, who contested Darwin on each and every one of the solutions he put forward, articulated no resistance to this "mystery of mysteries" description.[6]

What follows from the origin of species being declared a "mystery"? First, this mystery absorbs others, for science abounds with mysteries, and none more so than the science of generation and development. Why are two individuals needed to create one? What governs embryo formation? Is organisation the mere development of preformed structures? Such questions have been raised again and again, *ad infinitum*. In 1829, Cuvier described "the birth of organised beings" as the "greatest mystery of organic economy and of all nature".[7] In 1839, a physician, Henry Holland, multiplied the outstanding mysteries surrounding hereditary diseases and related physiological questions: "though the manner of their transmission be still a mystery, hidden in the same obscurity as the more general fact of the reproduction of the species, yet we are able to reason upon the effects, and to class them in certain relation to each other, and to the healthy and natural condition of the human frame"; the obscurity that shrouds causes should not prevent us from thinking about effects, from collecting "cases" and "new and wonderful instances".[8]

A scientific mystery is doubly troubling: it represents the missing answer to a question which may or may not, as far as current knowledge permits, be genuinely scientific. All mysteries emerge in the form of a question mark: not a solution, not even a problem. Faced with this mystery of origin, we may well begin by asking what actually *has* origin. As highlighted above, the meaning itself of "origin" is not clear and there are several ways of understanding it, as in the German concepts of *Entstehung* and *Ursprung*. Furthermore, it may be that the origin question just doesn't make sense when applied to the concept of species. If this were so, then it would have to be rethought by reorienting it either towards other taxonomical levels (origin of genera, origin of individual variations) or towards other fields (origin of man, origin of life).

As a concept, "origin" has questionable scientific legitimacy: whether it be pondered in a general sense or in specific relation to species, is it not inevitable that it will lead us away from scientific enquiry and into the sphere of religion, what Dupuis's *Origine des cultes* describes as "the satisfaction given to the soul's most mysterious needs"?[9] Mystery often draws on the lexicon of miracles and wonders. Huxley, for instance, speaks of the "development of a plant or of an animal from its embryo" as one of the "perennial miracles [Nature] offers for our inspection", "perhaps the most worthy of admiration".[10] Confronted with such a miracle, how should the naturalist act? By becoming more "conversant" with nature's operations, says Huxley. The deeper scientists' knowledge of nature, the more they will "wonder" at it and the less they will be "astonished". Study of nature must overcome initial astonishment, dumbstruck and admiring, in order to arrive at the effusive praise that sings of wonders, and only then should it move on to the questioning mode, discursive and inquisitive. Long before he lyrically portrayed origin as the mystery of mysteries, Darwin had read a book that "first gave [him] a wish to travel in remote countries, which was ulti-mately fulfilled by the voyage of the Beagle": this book was called *Wonders of the World*.[11]

Origin, outside of science?

Origins had haunted the eighteenth century: the origin of fables (Fontenelle), of languages (Rousseau, Monboddo), of inequality (Rousseau), of ideas and human knowledge (Hume, Condillac), of money (Smith), of government (Hume). The nineteenth century largely prolonged this same general line of questioning, so perhaps Darwin was just one current among many in a general movement, all eager to grasp the origin of everything, right down to the origin of seasons.[12]

What is it to investigate origins? Today, academic philosophy draws a contrast between an origin (absolute) and a beginning (relative, temporal), as well as between an origin (as static point) and a genesis (dynamic account). But actual usage is much fuzzier than this and the specific domain of origins is not quite so clearly defined. Further still, "origin" and "beginning" are subsumed as one under the notion of genesis, in the idea of a description of the beginning that is also the history of a production. Here, the origin is linked to the beginning, as the history of a primary term with its subsequent unfolding, as the depiction of an instant where, under a conflation of forces, something comes into being. That origin is in part to be understood as a beginning, and in part as a genesis, is an ambiguity created in no small way through translation of the biblical, Genesis account. Theologians asked what there was "in the beginning" or "at the origin"; "*in initio*" or "*in principio*".

In more than one sense, to seek the origin is to set oneself the task of retrac-ing a *genesis*, or, with all the ambiguity the term entails, a *history*: simultan-eously a description (the historian as witness, who has seen and therefore knows) and an account of a (re)constitution. It is to set the immobile into motion, to reveal the steps behind the progressive production of individuated entities and provide the reasons and causes of what, in appearance at least, is self-evident.

"Origin" lends itself to several interpretations. Either, when speaking of instances whose origin is given, it is used to empty those instances of their essence by marking the relative nature of their present state and showing that this is no more than the product of a deformation or disfiguration; or else, when pointing towards the primary state, origin leads us from the accidental nature of history to the absolute nature of the primary term, to the pure and immediate expression of a prototype in its very first instant. So "origin" can either be the blank page upon which the randomness of an event is set down and written, or it can be the primary term from which, via essential derivation, results the entire destiny of a form. In the first instance, "origin" invites us to make history, to pin down how transformations occur and individually emphasise the agents of trans-formation itself. In the second instance, "origin", wholly contained within the primitive point, forever refers us back to the cause or principle of this point source and to the laws of its development, proceeding just as though these were the only relevant questions.

If every originating formation is interpreted as an effect, then the existence of this effect implies that its origin has causes, and, should it call upon causes, then any account of origin will thereby necessarily entail reflections as to their agency: is their agency suspended following the instance of origin? And do these causes continue to produce effects infinitely, like in Spinoza's substance? In the latter case, there would be not *one* origin to be located at the beginning, before everything else, but rather each instance would be an origin, a point where the causes, endlessly producing their effects, would continue to produce, *to originate*, infinite effects. If the origin is an effect caught within a series of effects, then it is not to be understood as a starting point but as a process which can be attributed with a future and, undoubtedly, also an end.

The initial formulation of origin, its primary matrix, is to be found in sacred texts. Understood in this way, it loses all scientific legitimacy and becomes nothing more than an object of biblical hermeneutics (what did Moses write?) or poetic licence (what did Hesiod imagine?). In both of these eventualities, the scholar or philosopher can dispense with asking such questions. This seems to have been the general opinion held in and around 1850 by three major philosophical schools: one in Germany (post-Kantianism), another in France (positivism), and a third in Britain (inductivism). In his *Critique of Pure Reason*, Kant had disqualified all answers to the origin question by showing that the three primary objects of metaphysics—the soul, God, and the world—were entities inaccessible to the human mind. This created a mood of suspicion towards any attempt at answering the origin question. The same critical dimension is also present in the positivist tradition, where it fits in with the teleological vision of the human mind and its progress. The origin question is therefore condemned as being a regression, an investigation whereby the human mind turns away from natural causes in order to embrace supernatural ones; it brings about a transposition from science to superstition or religion, the abandonment of reason for imagination. In opposition to the obsession with the origin, positivism proclaimed its allegiance to "the school of facts". In England, the induction theoreticians placed origin beyond the sphere of science and ushered it towards the door onto religion. Even though not one of these three schools is reducible to any of the others, and even though they can be differentiated or opposed in many ways, they nevertheless seem to agree on their exclusion of "origin". Here again, their motivations may vary but they do all share what they see as an observation: that despite all of humanity's best efforts to penetrate certain great mysteries, the origin question has remained beyond man's reach. Thus, understanding the *Origin* is also a matter of seeing how Darwin's book reinvented both the theoretical origin framework and the debate over its scientific legitimacy.

Forbidding origin

Should the sciences turn a blind eye on the search for origins or not? For William Whewell, "the Origin of Man, and the Origin of Life upon the earth, were events of a different order from the common course of nature".[13] About both these matters and "the Origin of Language, of Law, of Social Relations, of Intellectual and Social and Moral Progress", he says: "though in all these characteristics of humanity we can trace a constant series of changes and movements, we can discern in them no evidence of a beginning homogeneous with the present order of changes".[14] Thus, *beginnings* (historical, knowable) can be set in contrast to an *origin* (radical, primal): where the beginning is homogeneous to the sequence it inaugurates, the origin is heterogeneous to it.

Of course, science hadn't actually taught that the origin of these phenomena must lie in a Creator's handiwork. But, Whewell tells us, it did absolutely admit of such a belief. Where origin is concerned, science steps back and leaves man free to believe. Here, to believe is to believe that the first step, that which brought certain realities into being, was a sudden leap, something the gradual changes we observe in nature today do not allow us to grasp. The cause that intervened at *the origin* of things would therefore not be of the same order as the causes that produce the modifications and transformations which surround us now. The origin would not be a beginning, a sudden arrival or interruption would have to be admitted; the ordinary would have to be left behind as the search shifted to the extraordinary, turning from Nature towards Providence. This is what the belief would be. According to Whewell, it would be to refuse man's origin among the apes, to refuse that language developed from a primal chattering or that reason was merely the result of conflicting appetites.

Condemnation for investigation into origins grew: for John Herschel, "to ascend to the origin of things, and speculate on the creation, is not the business of the natural philosopher"; for William Whewell, "all the lines of connexion stop short of a beginning explicable by natural causes; and the absence of any conceivable natural beginning leaves room for, and requires, a supernatural origin".[15] Darwin, with the title of his book, fully exposed himself to just such condemnation, even though, in other respects, he remained faithful to the spirit of this ban on origin inquiry; notably, for example, he excluded the origin of life and of higher mental capacities from his field of research.[16]

Even if the sciences must leave origin aside, the question of beginnings is nevertheless fundamental to what Whewell gathers together under the title of "palætiological sciences", "because they seek to ascend to an ancient condition of things by the study of present causes".[17] The paradigm of such sciences is Lyell's geology: it uses current causes, still in operation today, to explain past geological events, by referring only to uniform causes which have always operated with the same force as they do today (it pointedly avoids catastrophes and any notion of super-intense causes). Lyell's globally *steady-state* vision of the world makes impossible the determination of any kind of direction to history, whether this be towards some sort of progress or towards ultimate chaos. Lyell developed a geology wherein all periods are homogeneous and where changes are only oscillations. Thus, the principles of this geology open a way back up the chain of causes and effects, with the assurance that no event will rear its head to disrupt the order of causes known to us.

Whewell limits science to *beginnings* and refuses knowledge of origin as a primal point of departure:

> in all cases the path is lost in obscurity as it is traced backwards towards its starting point:—it becomes not only invisible, but unimaginable; it is not only an interruption, but an abyss, which interposes itself between us and any intelligible beginning of things.[18]

In contrast, Robert Chambers, in his *Vestiges of the Natural History of Creation*, throws back open the Pandora's box of origin: as long as the homogeneous field of current causes and their agency is maintained, a comfortable immanence is also maintained, ascending or descending, going from an effect to a cause or from a cause to an effect. But as soon as the supposition is made that science seeks altogether other (ancient) causes, then the epistemological framework of current and uniform causes so dear to Lyell is left behind in order to make room, within science itself, for other regimes of causality and lawfulness. In other words, science is freed from the constraint of homogeneous beginnings and the door is reopened to a scientific treatment of origin, of the radically *other*.

The whole ambiguity surrounding origin can therefore be found in this precise difference between Whewell and Chambers: for Chambers, we cannot account for origin through current causes—another class of causes, of a different register, must be called upon; for Whewell, such an approach is an impasse and, as such, "an impassable abyss separates us from the origin of things".[19]

In a dialog between Philocalos and Philalethes, newly converted to Darwin's theory, Julia Wedgwood depicts this issue of "origin" being out of the bounds of nature. If natural selection is a real law, then it is "only a name for a particular kind of Divine agency", and "God is not the less my Creator, if I am the result of this complex machinery", a result of natural selection as a secondary cause.[20]

History of science has also studied how Auguste Comte's thought may have indirectly constituted a significant element in Darwin's reflections on the nature of science. In reality, Darwin had most likely read only a review of Comte's *Cours de philosophie positive* that was published in the July 1838 edition of the *Edinburgh Review*.[21] From there to affirming that this reading, albeit at a crucial moment in the evolution of Darwin's theory, may have influenced the formulation of his own ideas might seem excessive. The review indicates the importance of *experimentation*, using individuals removed from their natural state and placed into artificial conditions—a point which may stir our attention in the direction of artificial selection and breeders' experiments. Furthermore, Comte, inspired by the Laplacian model, insists on the importance of numerical verification, something which may well be significant when we consider natural selection's debt to Malthus, who Darwin would go on to read the following October. But more significantly still, in this same review Comte is presented as *rejecting the notion that science is the search for causes*. Yet it certainly seems that Darwin, with his theory of natural selection, would precisely claim to provide the causes of certain phenomena. How, in that case, could Comte have been an epistemological model for him? In point of fact, everything depends on what is meant by "cause".

In mid-August, just after he had read the review of Comte's work, Darwin wrote:

> In my speculations, must not go back to first stock of all animals, but merely to classes where types exist, for if so, it will be necessary to show how the first eye is formed,—how one nerve becomes sensitive to light ... which is impossible.[22]

Darwin's theory works just like Laplace's. The latter explained the origin of the solar system by referring to an initial nebula, without obliging himself to go back any further. Darwin provides a principle, evolution by *descent with modifications*, but he does not claim to give the origin of life. And yet this is exactly what would be demanded of his theory. The review of Comte's work may well have comforted Darwin in the idea that one could in fact turn away from the notion of an origin of origin. Such an idea may have invited its still young reader to leave radical beginnings in the shadows and give himself over to the discovery of laws which are seized *in medias res*. As a case in point, let us recall that Darwin's "queer diagram" in Chapter IV—the only illustration of the *Origin*—begins, right at the very bottom, with, not one, but a *row* of dotted lines.

Origin (as distinct from beginning), therefore, appears to lie definitively outside of science: unthinkable, it seems to be inaccessible to our understanding because it is outside of time, outside of experience. In such an optic, it is but a mask for *ex nihilo* creation. With a view to avoiding this twofold trap, the 1850s saw the somewhat keen emergence of an alternative between the origin hypothesis (*ex nihilo* creation) and the idea that substance had always been. However, this second thesis ("permanentist" or "eternalist"), which denies any origin, is open to adoption in equal part by spiritualists and by materialists. It was a time when, in the USA for example, Theophilus Parsons was exemplifying the thesis of divine immanence in the following words: "the Causa causans must be always and incessantly a present cause, as present at one period of duration as at another, and always directly and universally operative".[23] Similarly, in France, Alfred Maury, showed how tenants of an *ex nihilo* creation lock themselves into a cycle: God, being only cause, exists only on condition of producing effects, or, if God is an act, then its existence can be understood only on the basis of some external material upon which it can exert its force, which is precisely from where the idea of a world/God co-eternity is deduced.[24] The *formation* of the world (its "origin" or its creation) is linked to its *conservation* (its persistence in being). From either vantage point, theological or naturalist, the notion of origin, viz. the origin of the creation, or the absolute beginning of everything, tends to be pushed out of the field of reason and science, abandoned to the realms of the imagination, to fable, to religion. It exits science and becomes a belief.

Judgements of the Darwinian intervention

The exclusion of origin, a foregone conclusion within positivism, affected how the *Origin* was received, particularly in France. As Émile Littré put it:

> all absolute questions, that is those questions whose affair is the origin and end of things, are beyond the sphere of human knowledge and, as a consequence, are no longer able to guide minds in their research, men in their conduct, or societies in their development. The origin of things, we were not there; the end of things, we are not there; thus we have no way of knowing either this origin or this end.[25]

Faced with this impossibility of attaining to the origin, men have turned to supernatural causes. From this point of view, the decision to abandon inquiry into origin emerges as the *sine qua non* condition of scientific progress. Positivism's claim was to have struck barren all research efforts in the direction of origin, and Paul Broca also signalled a point where "scientific research thus makes way, following the nature of minds, for philosophical doubt or for belief".[26]

Yet the Darwinian theory took upon itself to seize the origin problem and to run it back as far as it could, back to the origin of the first monad. But it too would rapidly meet its limits: "the cause of the first movement from unorganised matter to the state of organised matter, being beyond both explanation and hypothesis, lies outside the furthest limit of what can be known". Thus, the origin question brings us face to face with the question of what limits must be assigned to scientific knowledge and, continues Broca, "Darwin's hypothesis on the origin of species is not an essential part of anthropology but is inseparable from research into the origins of man, or rather the human type". A coalition formed of the positivists and Cuvier's disciples (including Broca) rose up in opposition to Lorenz Oken's *Naturphilosophie* and Étienne Geoffroy Saint-Hilaire's anatomical philosophy. All parties claimed to be "the school of facts", "the territory of fact, of empiricism, and of observation", and all agreed in declaring the origin of species question to be a *transcendent* one. Nevertheless, this question had not spoken its last word and would not lie low for long. For example, in 1869 Ludwig Büchner claimed that "all Geoffroy's opinions, then preconceived [i.e., in 1830]", were fully "justified today".[27]

The origin problem tends to be dealt with in different ways. For one thing, it is a little too easy to just push it away from science and into poetry, thereby believing the question to be done and dusted. Some, like Édouard Claparède, praised Darwin's boldness—"a man of heart who does not retreat before such a task. To remake the world! Or at the very least, to remake organised nature! What a crushing task!"[28] Ultimately, abandoning origin was perhaps not so much a sign of epistemological prudence as it was the mark of a cowardly age, the wily retreat of craven scholars, disciples of a science without ambition. Origin was headed to becoming a scientific Grail that only the worthiest of scholars, those bestowed with enough "heart", could ever take on. The legitimacy of the origin question is evaluated as though it were the object of moral qualities, rather than of simple epistemological criteria: you cannot have the first dimension without the other. At the same time, what short-sightedness to imagine that the inevitable could be evaded! "The mystery" comes into view as an inescapable moment with which, Claparède tells us, all scientific theories must collide:

> Whatever, indeed, may be the goal the physical and natural sciences set themselves in their research, they always come up against an irresolvable problem in the end, against a mystery. This problem, this mystery, is the origin and therefore the nature of force. Only before this problem is the scholar obliged to bow down and acknowledge his powerlessness. Mr. Darwin too had to encounter this mystery. It presented itself to him in the guise of the organising force. This is the unknown whose essence no mind will be able to examine. But if this force is to remain in itself forever mysterious, its action nevertheless can be studied, for forces are known only by their effects, and these effects manifest according to certain laws, the study of which is the very object of science. This is what Mr. Darwin has done.[29]

In this light, remarks criticising Darwin for chasing shadows are seen to be entirely unfair. In parallel, Eugène Dally, anthropologist and translator of T.H. Huxley, also became aware of these difficulties and set about reconsidering origin on the basis of a reinterpretation of positivism. Whatever Auguste Comte's definitive thought on the subject may have been, the warning recommending that science turn its back on the origin question could only serve as a guide to those undertaking specific research. For all others, according to Dally, the origin problem is necessary and "this is why an indulgent welcome, free of all systematic hostility, must be extended to every scientific attempt aimed at shedding light on a question whose solution, true or false, but accepted, has gained rule over the intellectual world".[30]

Dally, therefore, does not go along with the restrictive vision of our knowledge of origins, such as that presented by Broca. Instead, he proposes to restore the genuinely Comtean spirit. Consequently, in order "to combat an abused quotation" (by which he means Littré's, cited above), he recalls that for Littré "inaccessible does not mean empty or non-existent": "It is an ocean which crashes upon our shore and for which we have neither boat nor sail, though seeing it clearly is as salutary as it is wonderful".[31] In this, he sees a beautiful invitation to investigate the origin problem, not to push it away. If science was meant to forbid itself from all research into origins, it "would have abdicated in favour of just any cosmogony, even if, as we saw, it was of an absurdity that would derange the commonest of senses". The cosmogonies of his contemporary astronomers had been separated from any attempt at primordial explanation and from any conception which would take the universe as a whole for its object. Thus, they were fully aligned with the Kantian, positivist, and inductivist bans, and this is the sense in which the origin problem is scientific: "The universe, in its phenomenality, is a set of dynamic and plastic conversions, having neither beginning nor final term". Consequently, Dally proposed redefining the "meta-physical" level of questioning by distinguishing between the origin (knowable) and the cause (metaphysical). As a result, the epistemic ban finds itself either displaced or rolled back to another level:

> Of causes, we know nothing; a purely metaphysical notion carries no proof; it leans on no necessity, moreover it leads to no negation. Of origins, on the other hand, we know much and can affirm that, in the natural order, phenomena are but an undefined sequence of transformations. But keep oneself from concluding, because these forces transform themselves, that they are reducible to each other.[32]

Origins are taken to be beginnings, always locatable in the order of phenomena. We uphold a definitive relationship with origin, allowing us a certain way of knowing or conserving it as an object that can be legitimately scientific. However, we have no relationship with the *cause*, no more than we have access to what constitutes the *substratum* of phenomena (matter or spirit, or matter and force): "these terms refer to pure abstractions which [can] no longer play any role in science".

Now origin finds itself caught in an ambiguity. From one angle, evoking the biblical question of the world's beginnings, it touches on what gives rise to it; the supernatural, the ungraspable. But from another angle, it constitutes the first term in the sequence of all phenomena and, in this sense, is graspable. Thus, the origin takes form at the interface of what has no relationship to us (unknowable God) and the series of all phenomena (known through comparison). And so it *is* accessible, in a certain way.

Philosopher in short-clothes

The origin, just because it is not an object of demonstrative science, does not necessarily have to be an object of faith either. Asa Gray, for instance, notes: "The beginning of things must needs lie in obscurity, beyond the bounds of proof, though within those of conjecture or of analogical inference".[33] Hence, we find ourselves knocked back to the realm of hypothesis: what is not known by science can still constitute the object of conjecture. But, above all, the search for origin is not one and the same with the search for some ultimate term which would bring all such seeking to a formal conclusion. Rather, it is part of a vast unifying movement in the domain of the natural sciences. The origin question communicates the desire for

> reducing heterogeneous phenomena to a common cause or origin, in a manner quite analogous to that of the reduction of supposed independently originated species to a common ultimate origin—thus, and in various other ways, largely and legitimately extending the domain of secondary causes.[34]

This will raise the question of "how the diverse sorts of plants and animals came to be as they are and where they are". Such an investigation will be allowed to "[transcend] its powers only when all endeavors have failed": "Granting the origin to be supernatural, or miraculous even, will not arrest the inquiry".[35]

From a philosophical point of view, any genuine origin may be supernatural: when dealing with a given form, how can it be known whether it is truly the first form of all the mysteriously created primal forms? And when the philosopher arrives before this first term, isn't he/she confronted with the mystery or miracle of the event itself? Discovering the first term still leaves the primal miracle of becoming unaddressed. As Asa Gray recalls in jest, "to learn that the new-comer is the gift of God, far from lulling inquiry, only stimulates speculation as to how the precious gift was bestowed": "That questioning child is father to the man—is [the] philosopher in short-clothes".[36]

This "philosopher in short-clothes" rings true with the description Darwin gave of his young son Horace as "a prophetic type" of what naturalists would one day be. The day would come when, like little Horace, naturalists would dismiss thinking that "people *formerly* really believed that animals and plants never changed".[37]

The mysterious allurement of the forbidden

That origin became a focus of thought seems inevitable: the multiplication of hypotheses claiming to account for it is testament enough to that. Given this, what must be done is not to dismiss the question but rather to gather useful facts, facts liable to tip our judgement towards one or the other mechanism. What implication can be concluded from Darwin's insistent, almost ostentatious, manner of qualifying the origin as a "mystery"? Sociologists of science would have us believe that this was a classical homage to the scientific greats of his time. They would speak of authority. But, more precisely, how should origin be treated in light of his labelling it a "mystery"? Scientifically or mythologically? In a highly critical review published in 1860, Richard Simpson cut Darwin's book into two parts, "two elements, intimately blended": "One is the mythological conclusion" which "professes to give an account of the origin of man, of animals and of plants" and "the other is his accumulation and arrangement of scientific facts".[38]

And between the two, a chasm. Worse still, a flaw in the induction, particularly blameworthy since it is the very cornerstone of the book's irreligion. For Simpson, Darwin incarnates a new kind of infidel; following on from Voltaire's sarcasm and the mocking blasphemy of the libertine sceptics, Darwin represents the atheist who has the courage of his convictions, rooted in a calm and philosophical mind, free of all metaphysical ponderings but full of facts and examples which he believes authorise him to draw certain inductions. His crime? Having asserted that he could give a good *account* of the origin. Confronted with such discourse, what resources are available to those of true faith? Censorship being impossible, there remains disdainful silence, upholding the simplicity of faith.

Otherwise, there is argued dispute, the path Simpson chose. In this way, Darwin's principal crime in the eyes of the orthodoxy was to have forced theologians into elaborate reasoning around the mysteries of faith, mysteries which were otherwise perfectly clear in themselves.[39] "*Cognoscendo ignorari et ignorando cognosci*"; to be ignorant in knowing and to know in ignorance, this is the Augustinian lesson Simpson draws over the mysteries of religion. Much more than the ideas he advanced, Darwin's crime seems to have been opening the question of origin to scientific debate, plunging spirits and minds into aporias as torturous as those St. Augustine enmeshed himself in when meditating on time. Such aporias would not pass unnoticed: once the worm of reasoning has penetrated into the radiant fullness of mystery, doubt, like Proustian jealousy, spreads to every corner, and blind faith in love softly retreats. "Origin" appears to be an unsolvable and inevitable question. When the question persists in rearing its head, and when our answers are met only with repeated failure, is it time to give up or to risk new answers? Above all, to what extents must the quest be pushed? In this regard, the reception the *Origin* met with is exemplary of several misunderstandings. The demand imposed on Darwin was to provide answers to two questions he had not raised: the origin of life and the origin of man.

Notes

1 OS 1859, p. 1, *Var* 71 (#4).
2 See especially Notebook E, p. 59 (CDN p. 413). The reference is to a letter from J.F.W. Herschel to Charles Lyell, 20 February 1836. The letter had been published in Charles Babbage's *Bridgewater treatise*. See Cannon 1961.
3 Royer 1862, p. xv. Flammarion 1886, p. 110.
4 *Autobiography*, pp. 67–68.
5 Hooker 1853, p. viii; Jaeger 1860, p. 83; Gray 1877, p. 94.
6 Owen 1860a, p. 495.
7 Cuvier 1829, vol. 1, p. 15.
8 Holland 1839, p. 10.
9 Dupuis (1794) 1876.
10 Huxley 1893, vol. 2, p. 29.
11 *Autobiography*, p. 34.
12 Title of Mossman 1869.
13 Whewell 1846, Preface to the second edition, pp. 7–8.
14 *Ibid.*, pp. 7–8.
15 Herschel 1831, p. 38; William Whewell to Rev. Brown, 26 October 1863, in Todhunter 1876, vol. 2, p. 433.
16 *Origin* 1859, p. 207.
17 Whewell 1846, p. 9.
18 *Ibid.*, p. 166.
19 *Ibid.*, p. 9.
20 Wedgwood 1860–1861, p. 238.
21 Darwin read it between 7 and 12 August 1838. See Schweber 1977.
22 Notebook D, p. 21 (CDN, p. 337).
23 Parsons 1860, p. 10. Ospovat (1981) 1995 has provided an analysis of these Anglo-American debates.
24 Maury 1847. An analysis on the French context can be found in Hoquet 2009 (pp. 247–250).

25 Littré 1863, p. 107.
26 Broca, "Anthropologie" (1866), in Broca 1989, p. 34. On Broca, see Conry 1974, pp. 51–65 and Blanckaert 2009.
27 Büchner 1869, p. 19.
28 Claparède 1861, pp. 546–547.
29 *Ibid.*, p. 261.
30 Dally in Huxley 1868, p. 6.
31 Littré 1863, p. 519.
32 Dally in Huxley 1868, p. 15.
33 Gray 1877, p. 94.
34 *Ibid.*, p. 95.
35 *Ibid.*, p. 95.
36 *Ibid.*, p. 96.
37 Darwin to Gray, 23 November 1862, CCD 10 547.
38 Simpson 1860, vol. 12, pp. 363–365. On Simpson, see Lyon 1972.
39 Simpson 1860, p. 363.

8 "Originally breathed"

Or on the origin of life

Adam Sedgwick, the Cambridge geologist in whose footsteps the young Charles Darwin had rambled across Wales, rejected "the doctrines of spontaneous generation and transmutation of species, with all their train of monstrous consequences".[1] William Whewell, another professor at Cambridge, also vehemently criticised "the tenet of the transmutability of the species of organized beings" for the reason that it requires "certain additional laws" which "are still more inadmissible than the primary assumption of indefinite capacity of change": "a certain plastic character in the constitution of the animals, operated upon, for a long course of ages", and "the attempts which these animals made to attain objects which their previous organization did not place within their reach". This leads to the hypothesising of "certain monads or rough draughts, the primary *rudiments* of plants and animals", together with "a constant tendency to progressive improvement … which tendency is again perpetually modified and controlled by the force of external circumstances".[2]

As a matter of fact, the Oxbridge dons and most of the Victorian scientific elite had construed a theoretical framework intended to systematically transmute any view sympathetic to evolution into just another scandalous theory on the pile. In essence, the transmutation of species theory was made to appear as though fundamentally yoked with several scorned hypotheses: spontaneous generation, absence of design (*ergo* atheism), the bestialisation of man (the ape-man, of course, being the great fear of the Victorian age), and so on. But the transformation of species doctrine drags even more scandalous ideas along in tow.

With attacks on all fronts, the transmutation hypothesis found itself thrust up against its limits and obliged to clarify its assumptions. The creation hypothesis, on the other hand, had the answer to all its difficulties in one unique cause containing the source of all intelligibility and all intelligence. Between these two alternative scenarios, the Oxbridge dons chose the second since, having soaked the rational world in its principles and structure, the creation hypothesis presented itself garbed in the outer appearances of "knowledge". Darwin rejected this so-called knowledge which did nothing but "restate the facts": he went with the first scenario, opening up the Pandora's box marked "Origin". For committing this act, he found himself subject to all the torments and interrogations that came spilling towards him.

Whence progenitors and prototypes?

The *Origin* has often been read starting with the last page. Part of the book's last sentence drew the attention of many readers:

> There is grandeur in this view of life, with its several powers, having been originally breathed into a few forms or into one [...] [F]rom so simple a beginning endless forms most beautiful and most wonderful have been, and are being, evolved.

In the second edition, Darwin specified that that which had "breathed" was in fact "the Creator".[3]

This "breath" of life theme is also to be found a few pages earlier, in another important passage of Chapter XIV:

> Therefore I should infer from analogy that probably all the organic beings which have ever lived on this earth have descended from some one primordial form, into which life was first breathed.[4]

A third extract is also often quoted:

> I believe that animals have descended from at most only four or five progenitors, and plants from an equal or lesser number. Analogy would lead me one step further, namely, to the belief that all animals and plants have descended from some one prototype. But analogy may be a deceitful guide.[5]

Royer, in her translation of this passage, renders the Darwinian pairing "progenitors/prototype" with the opposition of "*types primitifs/prototype*".[6] In doing this, she reinforces the text's coherency around the origin question (the "*primitif*") but, by the same token, she also introduces a significant interpretative bias. The Darwinian term "progenitors" recurs frequently in the *Origin*, where it refers to ancestors, those who give birth to (and precede) an individual in a lineage. Therefore, "progenitors" have nothing specifically "typical" or "archetypical" about them, they are just the point from which *descent* proceeds. But who were the first progenitors? Darwin alludes to this in the final pages, and this is where many begin their reading of him.

The reason for this "upside down" approach to reading the *Origin* undoubtedly stems from the proliferation of commentary on the subject. In particular, 1859 is the year Félix Archimède Pouchet's *Hétérogénie, ou traité de la génération spontanée* was published in France, a point we will come back to.[7] This could be merely coincidence, but Pouchet's work does not only form a context within which Darwin would be received in France and abroad. More profoundly, (critical) examination of the *Origin* seems to lead necessarily to either the affirmation or refutation of spontaneous generation.[8] The *Origin* being understood as a theory of transmutation, it was expected, for the sake of its logical coherency, that it continues back, from the origin of species to the origin of "prototypes", taken to be primitive life forms, and, from there, to the origin of life. These three origins—species, prototypes, life—despite constituting distinct problems, were understood as irrevocably interrelated dimensions. They reconfigure the general coherency of the Darwinian system around its title: what must be known in order to claim genuine knowledge of the *origin of species?*

Darwin's success has often been linked to the *Origin*'s perceived radicalism. Readers primarily intent on isolating this curiosity-stirring radicalism would first turn to the final pages of the book, impatient to find its clearest expression. It is the case with the French botanist Antoine Laurent Apollinaire Fée, or with the English physician and ornithologist Charles Robert Bree.[9] A result of this headlong rush, which simply skips past reflections on natural selection and variation is that, instead of being perceived as an *originality*, the origin of species is judged as, in essence, nothing more than a variation on a canonical literary theme: a sole animal and a sole plant at the origin of all others. But if this were all the *Origin* boiled down to, their reasoning went, then Rétif de la Bretonne's fantastical *Austral Discoveries* had already contained the same proposal over a century before! Their assessment: what use was the *Origin* if its famous radicalism could be identified with and then shelved alongside such whimsical ancestry?

As Asa Gray notes, the concluding passages of the *Origin* run into each other and form a graded "wedge" allowing Darwin to pierce and then occupy the mind of his readers. When Darwin refers to *descent* from four or five progenitors, "we have the thin end of the wedge driven a little way". Then, when he refers to only one primordial form, "the wedge [is] driven home".[10] However, what Gray interprets as a facilitating element (the chock-like wedge that introduces Darwin's ideas into the minds of his readers) can also represent the axe-like wedge driven into a block of wood to split it in two. The stark contrast of this "wedge" ambiguity underpins many of the work's critiques.

Darwin's most fervent followers criticise his excess of prudence, if not cowardice. On the side of the popular materialists, Ludwig Büchner was particularly harsh:

> Darwin—and this is a defective side of his doctrine—either didn't have the courage, or else didn't have the logic, to pursue and carry his idea of a common origin for all beings to its final and extreme consequences. He stops at four or five *primordial forms* or *root couples* for the animal kingdom, as many for the vegetable kingdom, and he admits that, at the origin, in the mists of time, these types were called into existence by the creator.[11]

Others, by contrast, accuse him of being excessively bold. Flourens, for example, said:

> I see, first of all, that his system has no beginning. The obligatory beginning of every system, which integrally constructs beings, is *spontaneous generation*.

Whether we try to fend it off or not: all systems of this sort begin through *spontaneous generation* or else lead to it.

Here, Darwin is believed to have climbed back from ancestor to ancestor, without end:

> In natural history, there are only two possible origins: either *spontaneous generation*, or the hand of God. Choose. Mr. Darwin writes a book on the *origin of species*, and, in this book, what lacks is, precisely, the origin of species.[12]

Flourens mocks Darwin pitilessly: he arrived too late, into a century where spontaneous generation was no longer a credible belief. Lamarck, in comparison, was happy to find refuge for his transformism in the faulty doctrine of spontaneous generation. But now that "spontaneous generation is not", only one of the two possible origins for all living beings remains: the hand of God.

The debate relating the *Origin* to spontaneous generation raged well beyond just France's borders, and well beyond Pouchet's book. The same debate reminds us of the existence, notably in Germany, of a deep-seated "autochthonous" movement, according to which life had sprung from the depths of the earth.[13]

This movement constituted a framework for examining the Darwinian theory as well as a touchstone for its evaluation. Hermann Burmeister, notably, published his *Geschichte der Schöpfung* in 1843, a work which ran to several editions. The book, never translated into English, includes a defence of *generatio originaria*. The fact that this phenomenon is no longer observed today, that its principal bases no longer stand, does nothing to advance the question of life's first origin:

> Thus, not wishing to take refuge in either miracles or mysteries, we must concede that the origin (*die Entstehung*) of the first organic creatures on Earth was in the free, procreative ability of Matter itself, and the reason why this procreative ability no longer persists is to be deduced from the general laws of Nature being set to follow only necessity, not superfluity.[14]

The text sparked violent reactions and Burmeister was accused of

> gratifying irreligious passions by doing his utmost to erase the role of the Creator from creation [...] And so, in place of the rational dogma of God's creation, is substituted the most outrageous of miracles and mysteries: the production of life by death and of order by disorder![15]

A non-sexual generation

When philosopher Paul Janet took on the question of spontaneous generation in 1864 he worked only from German texts, without citing Darwin. It is "one of the

most obscure problems of human science, before which a circumspect philosophy would wish always to remain silent rather than propose hypotheses so
onerous to verify".[16] Here, the problem of spontaneous generation brings us back
to the plurality in modes of generation, a consideration that was central to the
interpretation of the theory of descent and to the fact that not all reproduction is
sexual: "experience teaches us that there are greatly varied modes of generation;
why should not one of these modes, at the lowest degree of animality, be heterogeny?" Against this objection, Janet notes that the scientific movement of the
nineteenth century was running contrary to the Enlightenment scientific movement: where Charles Bonnet and Abraham Trembley had worked to moderate
sex and gender, discovering, one working with aphids and the other with polyps,
cases of generation through budding, the nineteenth century seemed intent on
finding the division of the sexes wherever it could, expanding the realm of sexual
reproduction whose very universality the Enlightenment had questioned.[17]

Spontaneous generation questions the universality of sex as a mode of reproduction. This point also struck Gustav Jaeger, one of the principal defenders of
Darwinian theory in Austria. Every animal owes its existence to a birth; is it not
then unthinkable that anything could have been "produced [*entstehen*]" in some
other manner? This seems to constitute an unacceptable exception to the grand
laws of nature which have occupied the greatest thinkers since Plato. If the status
of sex as a constant of nature is thrown into doubt, then all human knowledge
may be nothing more than "vain mind contortions". Sexual reproduction is therefore the gauge of nature's constancy, constituting, according to Jaeger, the very
heart of the naturalist enterprise. This constancy goes unchallenged in the physical sciences (astronomy, geology), where the laws of nature appear to be valid
across every epoch and where the term "revolution" refers only to the eternal
recurrence of the same. However, the dominion of constancy is shaken in the
realm of the organic sciences: certainly, today, there is no more sign of *Generatio aequivoca*. However, it is at least possible that in far-flung times animals
"originated [*entstanden*]" neither from buds nor seeds. It is on this exact and
crucial point, Jaeger tells us, that Darwin opened the door to explanation. The
challenging of sex is not exclusive to Darwin's views; many had observed the
successive appearance of infusoria as they reproduced by division, of plants
through budding, of individuals multiplying through spores rather than grains or
seeds, and so on. But Darwin's system finally cleared the way towards explaining the reason behind this plurality of reproductive modes by showing how each
animal form splits away from the preceding ones through the accumulation of
individual differences in a given direction.[18] Jaeger sets the roots of Darwin's
views into the *Generatio aequivoca* context in order to cultivate scientific study
into the evolution of sex. Above all, Darwin introduced a scientific approach to
these questions; his views shoot down ambiguities and produce explanatory
mechanisms. In fact, it is better to avoid a term like *generatio aequivoca* and the
element of mystery it suggests.

The *Origin* was equally well received by believers in a "history of nature",
such as Heinrich Georg Bronn. In his *Handbuch einer Geschichte der Natur*,

Bronn called for a vision of nature, both organic and inorganic, to be established upon "historico-developmental foundations [*entwicklungsgeschichtlicher Grund-lage*]", "in the same way that we have a history of peoples and of states".[19] His goal was to conceive of "the whole of nature as one big organism" and to study "the interactions of each member of that organism ... as well as the forces which preside over their principle and their lawfulness". A prizewinner of the Académie des Sciences in Paris in 1856, Bronn's findings went on to be widely circulated. In England, even before Darwin, they provided the basis for spontaneous generation to be placed within the conceptual framework of its connections to the origin of species and the question of "organic creation".[20] In his reading of the *Origin*, Bronn moved aside the origin of species question and took a step beyond it with his radical decision to investigate the origin of life and the origin of variations.[21] Interpreting *Origin* as *Entstehung*, Bronn proposed an etiological programme to Darwin—*Ursache* rather than *Ursprung*—in which natural selection (reinterpreted, as we saw earlier, as *Wahl der Lebens-Weise*) plays a decisive role as "the richest and most general cause".

Bronn analysed the logic of the Darwinian system: carried to its logical conclusion, it implies that "all plants and all animals stem from [*herrühren*] a single prototype" and, therefore, that all forms can be connected back to four or five progenitors (*Stamm-Individuen*). Once the existence of several "primordial types [*Urtypen*]" has been admitted (whether this be ten, five, three, or even two), a creation is necessary. But if there were only a single, unique prototype, then the organic world could have its origin in a primordial goo, such as Priestley's "green matter". Experiments on spontaneous generation seemed, to Bronn, to be crucial for the status of Darwin's theory, not to mention the fact that they would strike right at the problem of sex in nature. If "organism species" could be spawned, in certain conditions, without the need for organic germs, then Darwin's theory "would receive the greatest support possible, in the shortest possible time", on condition that the "direct origin [*direkten Entstehung*]" of organic matter from some inorganic elements be demonstrated. Conversely, for as long as these two points (generation without germs, production of organic from inorganic) were not established, the Darwinian theory would always have need of recourse to a "creative force". And once *one* act of creation has been admitted, why not admit thousands of them? Likewise, for Bronn, anyone adopting Darwin's views is backed into admitting two corollary hypotheses: not just the idea of progressive development but also, inevitably, the idea of "things having a primary beginning". Bronn asks Darwin's *Origin* to prove the possibility of using inorganic matter to produce organic matter "endowed with a cellular structure" and, further still, to prove that from this matter "the seeds and eggs of the inferior organism species" may be obtained. If science were to produce such results (which does not seem beyond all possibility), then it could "cease to have recourse to personal acts of creation which have no place inside the laws of nature". Bronn did not attack Darwinian theory for not being demonstrable; as it stood, proofs and refutations were not forthcoming, and, in the meantime, there were those who believed in it and those who did not.[22] However, he accused

Darwinian theory of barely carrying our understanding forward at all: it remains "all the more implausible for the reason that it does not lead us to the solution of the great problem of creation". It leaves the only problem that really matters outside of its field, the origin of life, untouched. The intervention of even one creation is a fatal sprain for the system's general construction; the wedge stabilising the system is ever at risk of becoming an edge which, if fully jammed in, could bring the whole system tumbling down. Thus, it is not at all incidental that the broader origin of life question should cross paths with Darwin's own.

Darwin's is not the only theory to have known such inflection; it can be seen that the same processes were also applied to the domains of cellular theory and organic chemistry. Ultimately, for Bronn, the question of the origin of living forms ties in, first, with the theory of descent through the omnipresence of the cell in the make-up of organisms and, second, with the synthesis of organic compounds from inert elements.[23] Spontaneous generation must be specified as part of the problem of producing a *cellular* structure, of producing organic from inorganic, and also as part of the possibility that this path may produce *species* or, in other words, organisms which can reproduce through germs. Leaning on cellular theory, Bronn asserted that there is nothing surprising about the development of all specific forms from some few prototypical ones: we see it happening with our own eyes, since a unique being is manufactured during the embryonic and foetal life phases, beginning from just a single cell.[24] But this merely makes the problem more acute and its solution more urgent: "there can be no light shed for as long as the origin of life is not explained".

How many prototypes?

Darwin wanted to avoid the question of the origin of life. Adopting a very Newtonian stance, Darwin stated, "surely it is worth while to attempt to follow out the action of Electricity, though we know not what electricity is".[25] Howard Gruber states Darwin's position on the matter like this: "by avoiding the issue of the origin of life, the *Origin of Species* gains in simplicity what it loses in scope".[26] But in fact the issue is more complex, as Jeff Wallace has emphasised; the fact that readers of the *Origin* obsessively dragged Darwin back to the question of the primitive prototypes "is also a useful indication that Darwin could not draw up his own interpretative parameters for the *Origin*".[27] Darwin finds himself subjected to a forced march, marshalled independently of his actual words.

Thomas Vernon Wollaston, for instance, in his review of the *Origin*, discusses the issue of a definite number of prototypes being created:

> To our mind, the wonder consists in the act *at all*, and not in the number of times that it may have been repeated; for a Being that *can create* may surely do so as often as He pleases; and we have no right therefore to limit that act,—at any rate on the question of its *probability*; for, if we admit that it has been exerted so much as once, there is no a priori reason why it should

have been a million times repeated, or why, if He had so willed it, it might not, at some period or other, have been in constant operation.[28]

T.H. Huxley tries to move away from the topic of creation as he observes that Darwin's " 'transmutation' hypothesis is perfectly consistent either with the conception of a special creation of the primitive germ, or with the supposition of its having arisen, as a modification of inorganic matter, by natural causes".[29] Here, Huxley is asserting a theoretical polymorphism to the *Origin*, showing that it should not necessarily start with an act of creation.

Attempts to keep Darwin separate from these questions have been rare. In this sense, Auguste Laugel is unique: he considered the fact that there is no demonstration of spontaneous generation as an open door to the idea of a creator. Admittedly, "creation's thread hangs, in Mr. Darwin's theory, from some unknown thing". But must he be criticised for this? The *Origin* does indeed constitute a first effort at banishing the idea of creation, or at least forcing its retreat, yet, from a religious point of view, it does also have the advantage of being a reduction to a single "thread", whereas the theory of discontinuous creations had torn this thread "into a multitude of parts".[30]

Armand de Quatrefages used prototypes to measure the respective advantages and disadvantages of Lamarck's and Darwin's theories. The former supposed the constant production of elementary organisms which then evolve towards more elaborate stages; the latter, manifesting a certain "wise restraint", supposed initial prototypes which he then had evolve through slow derivation. The Lamarckian system is in absolute need of ongoing spontaneous generation, whereas Darwin quarantines this question to the sole origin of prototypes, with respect to which he exercises a restraint whose implications are deserving of analysis. Darwin does not go as far back as Lamarck: in other words, he doesn't in any way attempt to explain the existence of his prototype, a reservation he has been universally chastised for: "He was criticised for leaving his theory incomplete, for not holding fast to what the title of his book had promised, backing away from the question of first origin". Quatrefages, on the other hand, praised this attitude: "Every man has the right to personally decide on the limits where his knowledge shall end. Furthermore, Darwin's declaration concerning spontaneous generation abounds with good measure and sense. Here, his is the language of the genuine savant".[31] Quatrefages is not asserting that spontaneous generation is impossible. More so, his remarks stress that it is neither refuted nor demonstrated, if not in fact being irrefutable and indemonstrable. On such an issue, it is better to admit ignorance. In contrast, when Lamarck's theory is confronted with the primordial blind spot of spontaneous generation, it loses its very foundation: it emerges lacking the starting point of all its transformations. Darwin, however, "by refusing to explain the origin of life, by taking the living being as a primitive fact, thereby escapes all such difficulty". Of course, removing spontaneous generation to outside the domain of natural phenomena does not therefore resolve the mystery. But the strength of Darwin's system is that it allows the unknown to roam free, something that makes it compatible with a variety of hypotheses.

Here, spontaneous generation is specifically related to the origin of prototypes: the general, and quite abstract, question of life's appearance on earth
transforms itself into the more specific question of the "forms of life". The
source of organic (as opposed to inorganic) matter is reformulated as the origin
of elementary types or body plans from which evolution works. Does this mean
that the vital cause is to have created only one sole archetype before ceasing its
action for all eternity? We would then be dealing with a unique case of a natural
cause acting only once. Quatrefages points out the singular nature of such a
hypothesis: is not the specificity of a natural cause, as opposed to a miraculous
one, that it arises steadily, constantly, repeatedly? In such a perspective, we can
no longer make sense of life appearing only once on earth. If life appeared in a
non-miraculous manner, then it can't have happened just once, but rather many
times. This, in turn, calls for an investigation into the number of individuals initially representing the prototype.

This question gave rise to a polemic among French readers of Darwin: Clémence Royer, denying the single ancestor, supported the view of multiple
primary organisms produced by natural causes;[32] Eugène Dally opposed the idea
that the earth might have functioned as a miraculous and incomprehensible "universal matrix". Royer's position, that the earth had produced several seeds at a
certain point of its history, turns out to fit quite nicely with the intervention of
some supernatural power in nature, "intervening here and there in its creation, in
accordance with its mysterious fancies".[33] Likewise, opponents of Darwin, such
as Hyacinthe de Valroger, commented with sarcasm on the origin of these prototypes and all such materialist speculations around the origin of life:

> Following Mr. Darwin, the Creator must have created as little as possible.
> Miss Cl. Royer, on the contrary, affirms that *Nature*, the real creator, *never
> saves on seeds* in its *creative generosity*; and so, one must believe in the
> *infinite multiplicity* of primitive seeds.[34]

Openly making fun of this "infinite" seed generation, Valroger concludes:
"When one denies the miracle of God's creation, one is condemned to dreaming
up impossible miracles". Royer's polygenism, invoked in order to naturalise the
miraculous origin of the first seeds, can only lead to a multiplication of miracles
the result of which is to render the whole explanation suspect.

Consequently, in terms of both the origin of seeds and the number of prototypes (two intimately interwoven questions), Darwin's ideas began to be overtaken, which in itself can be understood as either their accomplishment or their
relegation. The origin of prototypes authorises three kinds of transformism:
polygenic (Royer), oligogenic (Darwin), or monogenic. The difference
between these three kinds boils down to the number of seeds initially supposed.[35] The theory that supposes life to have a natural origin, making natural
products of living beings, leads to the supposition of multiple centres of organisation: the very logic of the *Origin* would then lead to a polygenism.[36] In
Broca's words:

The fundamental notion of current transformism, that is, that living beings are natural products, would seem to me to lead logically to the idea of multiple origins, multiple in time, multiple in space, multiple also in their primordial forms—in a word, to a polygenic transformism.

This polygenic position arises naturally from the belief that natural forces spontaneously produce both the organisation of the living and the emergence of types, or of seeds, being mutually identical. But if the types have been produced several times over in the same way, then it must be supposed that there exist resemblances in nature which cannot be explained through descent. Polygenism, the implication of a naturalist reflection on the origin of life or of prototypes, leads to a weakening of the arguments in favour of the origin of species (resemblance as a sign of derivation or transmutation). Once the sources are seen to be many, the polygenic position is paradoxically reduced to a refusal of descent: the structure analogy can no longer be considered as sufficient proof of a common descent. The origin of germs resonates louder than the origin of species. Choosing a monogenic hypothesis seems the equivalent of a unique and miraculous creation but has the advantage of assuring filiations between resemblant forms. If it is then supposed that this origin was natural, it must also be supposed that it occurred several times—a few times at least, but more plausibly a great number of times. Extending the naturalist way of reasoning leads to a division between mono-, oligo-, or polygenism, and whatever solution is retained must be questioned according to the relations between (natural) causation and (miraculous) creation.

In contrast to Darwin's guilty offhandedness or wise reserve (the latter of which tends to give the impression, in Chapter XIV, that the question of prototypes is a secondary point), the work undertaken by these interpreters of his work instead highlights the most profound implications of this particular question.

Two men of genius

Related to the above considerations, comparison of Darwin and Pouchet was a widespread commonplace, as evidenced by some few lines which appeared in *Blackwood's Edinburgh Magazine* for February 1861:

> While Mr Darwin's *Origin of Species* has been occupying the scientific and semi-scientific circles of England, a similar agitation has been excited among the scientific circles of France, by the Memoirs presented to the Academy, as well as by his more elaborate Treatise, in which M. Pouchet has proclaimed the doctrine of Spontaneous Generation. The work of Mr Darwin differs greatly from that of M. Pouchet in ability, in novelty and in philosophic spirit; but both works are calculated to excite polemical passions, and both accordingly bring into painful relief the very imperfect condition of our scientific culture.

Then, a few lines below:

> there is no practical result to be anticipated from the Development Hypothesis, or from that of Spontaneous Generation; but there is a speculative result of no little interest, inasmuch as these hypotheses are thought to affect certain meta-physical views of Creation, supposed to involve important consequences.[37]

Thus, the two books, Darwin's *Origin* and Pouchet's *Hétérogénie*, are comparable (and compared) with respect to their methodological status (both are hypothetical), their impact (both are controversial), and their implications (both are metaphysical).

Darwin's book, extended to its logical conclusion, inevitably leads to Pouchet's: the theory of descent with modification implicitly moves towards the theory of spontaneous generation. Despite the stiff condemnations it was subjected to, and even after Pouchet's refutation at the hands of Pasteur, the thesis of spontaneous generation was not such a shocking one in the late 1860s, and Pouchet's *Hétérogénie* was quoted favourably by Richard Owen: "Besides the superiority in fact and argument, Pasteur, like Cuvier, had the advantage of subserving the prepossessions of the 'party of order' and the needs of theology".[38] Owen maintains that Pouchet, relative to the theory of the origin of monads, played an analogous role to that maintained by Étienne Geoffroy Saint-Hilaire regarding the origin of species. Both figure among those intellects who see truth behind veils of obscurity and persevere in affirming their theory, despite all contrary proof: "*Eppur si muove*", comments Owen, baptising them into the Galilean tradition of great scientific counter-evidences.

Consequently, the attacks against Pouchet were not left unanswered. In a materialist defence, Darius-C. Rossi jointly defended the *Origin of Species* and spontaneous generation, or "heterogeny": Darwin and Pouchet are compared as two geniuses who shook common opinion. If forms do change,

> where do the first manifestations of life begin to break through? Where do the first phenomena of mutability arise? Which power sets them to work? Where do their efforts end? All so many unfathomable mysteries. But if impenetrable laws do govern the evolution of life, then their results cannot be contested!

Heterogeny and descent with modification

> obviously exist; but what remains to be done, and this is the supreme effort, I repeat, is to trace the limit where both powers end; whosoever is genius enough, to employ Linné's expression, *erit mihi magnus Apollo* [shall be a great Apollo].[39]

Thus, it is befitting to limit the forces of spontaneous creation and of mediate derivation, as well as the powers of the origin of germs and the origin of species.

Rossi reminds us that "Mr. Darwin does not claim to demonstrate the primal origin of mental faculties; but he has sufficiently proven that they can change, just as there may be modifications in physical organisation, without, of course, any need for simultaneity". Hence, "all is not explained, it would seem; yes, Darwin would not disagree: he himself recognises, with rare steadfastness, the impossibility of the task". But for Rossi, belief in a creator is just as lacking in proof as the transformation hypothesis.

Above all, Rossi would have wished for Darwin to tell us more about the origin of his prototypes, and he storms:

> But who then created the Darwinian prototype? It clearly came into the world ... Shh! Darwin has the extraordinary good sense not to mention it, a *true savant*, and this silence ... brings enormous relief to the chest of our academician.[40]

Whereas Quatrefages praises Darwin for his reserve, Rossi, taking a leaf from Royer's book, formulates an opposition between irreconcilable systems: one stands either for Pouchet's heterogeny or for Pasteur's panspermia; there is no third term. Thus, it appears that although readers of the *Origin* may mention or recall ideas that are original to the Darwinian system (struggle for existence, natural selection), the fundamental framework of Darwinian thought finds itself transposed to a general schema of derivation from prototype: the Darwinian question of the *Entstehung* is interpreted as an *Ursprung*. In such an approach, "Darwinism" is relegated to being little more than a rival theory to Lamarckism, located entirely on the same level of explanation. By shifting Darwin's work from the origin of species to the origin of prototypes, his readers reformulated his intention and constantly held him up against the logical necessity of his own system, whether this was to extract an argument in favour of spontaneous generation or, on the contrary, to refute it in the name of Pasteurism. Through this, Darwin's views are assimilated into the great expanse of theories on the transmutation of prototypes.

In contrast to Lamarck's temerity, Darwin shows himself to be circumspect. But because of this, he was criticised for the name of his work: *On the Origin of Species*. As Puech, a doctor from Montpellier, put it: "while he does show us where the species is going and how it is currently produced, nowhere does he tell us where it comes from". As a result, Darwin's reserve seems a simple "stunt", a ruse which does not quite work: "Obviously, [Darwin] believes in spontaneous generation; but then why does he not state so categorically? Why not make use of a less equivocal wording?"[41] Spontaneous generation and primitive prototype would be alone in giving a clear answer to the very question Darwin posed: the question of "the origin".

Admittedly, all of these debates *around* the *Origin* rather than directly *on* its content, did not occur without having some impact on certain of Darwin's formulations and on the manner in which he was read.

Introducing the breath of the Creator was one such important effect. This variant has been much criticised and, commenting it in 1863, Darwin said:

Your reviewer sneers with justice at my use of the "Pentateuchal terms", "of one primordial form into which life was first breathed": in a purely scientific work I ought perhaps not to have used such terms; but they well serve to confess that our ignorance is as profound on the origin of life as on the origin of force or matter.[42]

So, like "chance" variations, this breath metaphor for the origin of life is seen also to serve only as a mask that veils another aspect of our ignorance. The reserve Darwin showed with respect to this Creator reference provoked the following withering commentary from Hyacinthe de Valroger:

In 1863, he bashfully excused himself for his employ of this biblical metaphor! He should rather have recognised that he had improperly taken up a metaphor *Genesis* had intended only to represent the creation of the human soul and then applied it to the production of the lowest of all creatures in the organic world. But the memory of the holy text must no doubt have been erased from his mind by his studies on pigeons, barnacles, and natural selection.[43]

Valroger relentlessly criticises the language Darwin employs in talking about God: "most frequently, his language attributes the Creator with the processes, methods, and habits of horticulturists, breeders, masters of stud farms, the members of a *Pigeon's club!*" He is particularly critical of Darwin's use of expressions where the terms "intelligent power" and "natural selection" are put together or near each other.

This shifting of terrain, from the origin of species to the origin of prototypes, provokes a complete reconfiguration of the text, where God becomes the true origin and the very term "natural selection", by its proximity to "artificial selection", appears base and profane.

Did natural selection operate on the original plasma? (Bastian)

Henry Charlton Bastian also cast the *Origin of Species* back towards the origin of life, but in order to conceive of the "progress" between forms and to see if natural selection applied to primitive forms.[44] Are the forces of Darwinian evolution (natural selection first among them) capable of assuring the progressive development which led from "inferior" organisms to "superior" organisms? From the case of "inferior organisms", Bastian inquires into "the grand question of the origin of life". Everything seemingly began from a vast "plasma [*plexus*]" made up of "complexly interrelated" individuals, forming the infusorial or cryptogamic life which he calls the "ephemeromorphs", comparable to crystals. From this point, it is necessary to know whether or not natural selection operated on the elementary and infra-specific level of this primordial plasma. From the Darwinian point of view, it seems that natural selection must apply at all levels, including the initial plasma state. Conversely, for Bastian, natural selection can have no influence on the ephemeromorph level; neither do the laws of inheritance apply, leaving only a scenario of simple, variable, and irregular forms, where only "laws of polarity" exercise a determinant role in the production of structure. Bastian therefore responds in the negative: selection constitutes "these subtle and slower modifying agents" and does not apply to the first stages of life; it can only intervene once life has become more complex, through a movement of "ascending development" whose mechanism Bastian does not explain.

Bastian's response here may surprise some. However, he does take care to ground his response in two separate passages from the *Origin*. One, taken from the introduction, posits the grand principles which command the natural selection of individual variations; the other, taken from the summary of Chapter IV, equates the differences between varieties of the same species and the differences between species in a genus.[45] Bastian concludes from this that natural selection cannot act on the elementary level of these "ephemeromorphs" because, in order to enter into action, it would need something like "homogenesis" (same produces same, or everything comes from an egg) and the laws of inheritance. In other words, *it appears to him that natural selection only applies once something like species has been established*, and Bastian, interested in a *pre-species* state, feels obliged to make room for other operational forces.

Bastian is not "anti-Darwinian". The starting point of his reflections is natural selection and its definition; he fully recognises its functions and attempts to evaluate its role, somewhere between dividing blade and creator. For him, natural selection operates a "production of variation" which is confined to the stage where the laws of inheritance have already been established. This is why he considers species (forms which reproduce themselves consistently) to be necessary in order for natural selection to act. Bastian clearly distinguishes between the conserving and producing functions of natural selection: "natural selection as a maintainer of already established forms, and natural selection as producer of new forms".[46] Darwin was the first to have perceived this second case: "To him we owe the discovery that Natural Selection is capable of *producing* fitness between organisms and the circumstances".[47] The Darwinian contribution was to have shown that natural selection is not only a sorting device but also a device by which *fitness* is produced; that natural selection does not only act as a dividing blade but as "an ever-acting cause of divergence amongst organic forms". Nevertheless, did Darwin really account for the gap opened up by the shift in philosophy implied by this new conception of natural selection? In the 1868 *Variation*, he declares that natural selection works from a base of variability up, or, in his own words:

> natural selection depends on the survival under various and complex circumstances of the best-fitted individuals, but has no relation whatever to the primary cause of any modification of structure.

In the 1875 version, this same text becomes:

> selection depends on the preservation by man of certain individuals, or on their survival under various and complex natural circumstances, and has no relation whatever to the primary cause of each particular variation.[48]

Several significant shifts are implied by the move from the first to the second of these two versions. Darwin introduces a broad, unspecified general "selection" (which includes artificial selection) to replace the "natural selection" present in the initial version. The second version also involves placing preservation and survival on the same level, as well as shifting from "modification of structure" to the general question of the origin of variations. According to Bastian, these modifications show that Darwin did not properly differentiate between these two joint roles of natural selection, as both the dividing blade that merely conserves and also the genuine producer of novelty. Otherwise, why would Darwin have reduced the productive dimension of natural selection to the mere maintenance of favourable variations? Here, Bastian is highlighting Darwin's neglect of other causes that produce adaptation due to the exclusive privilege he grants to natural selection, and this despite the fact that he never properly distinguished the latter from its negative and purely eliminative function. Darwin's failing, in Bastian's mind, was in taking one particular cause of novelty production to be the general agent at the origin of all production.

By scaling down the agency of natural selection, Bastian believed himself to be providing a fair re-establishment of the facts and an account of the other factors at work behind the production of novelty. He did not, however, seem to notice the double level of variation production and *fitness* production stemming from the action of natural selection upon these variations. Far from understanding this as a conceptual clarification, such a distinction between two levels of variations (primary variability and the appearance of new fit forms) seems to him, on the contrary, to be based on a primary confusion between two meanings of natural selection (dividing blade and producer). To eliminate this confusion, Bastian wanted to push the influence of exterior circumstances upon species back to the fore and, towards this aim, he identified three principal causes for the production of new variations: exterior forces which modify the internal balance; a functional modification which produces an alteration of structure (indirect influence of conditions), and; the perpetuation, over generations, of useful modifications. The first two categories can be seen as starting points for natural selection. But, more generally speaking, Bastian backed Herbert Spencer in affirming that these causes act directly as producers of variations or adaptations: "So that they must stand side by side with natural selection, if not as co-equals, yet as occupying marked and important positions".[49] Bastian's text provides us with a clear illustration of how problems interpreting the concept of natural selection are echoed in the role that is attributed to it: the agent (natural selection) is constantly related back to the material (variability) and Darwin finds himself constantly dragged from the origin of species to the origin of variations.

Notes

1 Sedgwick 1831, p. 305.
2 Whewell 1857, vol. 3, pp. 479–481.
3 *Origin* 1859, p. 490, *Var* 759, #270. For recent examples of fascination with "endless forms", see Howard and Berlocher 1998 or Donald and Munro 2009.
4 *Origin* 1859, p. 484, *Var* 753 (#220).
5 *Origin* 1859, p. 484, *Var* 752 (#215–217) (emphasis added).
6 Royer 1862, p. 669.
7 Pouchet 1859.
8 See Conry 1974, pp. 71–72, and Fry 2013.
9 Cf. Fée 1864, p. 1; Bree 1860, p. 1 begins with quotations excerpted from a book, entitled *"On the Origin and Variation of Species"* (*sic*).
10 Gray 1877, p. 105.
11 Büchner 1869, pp. 67–68.
12 Flourens 1864, p. 68.
13 On the issue of autochthony, see Rupke 2005.
14 Burmeister 1851, p. 325.
15 Valroger 1873, p. 48; James 1877, p. 73.
16 Janet 1864, p. 92.
17 See for instance, Churchill 1979.
18 Jaeger 1860, p. 87.
19 Bronn 1841. See Gliboff 2007 and 2009.
20 Cf. *Quarterly J. Geological Soc.*, no. 57, February 1859 and Powell 1859, pp. 463–469. On Powell, see Corsi 1988a and 1988b.
21 Darwin to Bronn, 5 October 1860, and to Lyell, 8 October 1860, CCD 8 resp. 408 and 421.
22 Bronn 1860b, p. 113.
23 The point is remarked by Büchner 1869, p. 69 and p. 81.
24 Cf. for example Bronn 1860a, p. 514 and Büchner 1869, p. 67.
25 Darwin to Bronn, 5 October 1860, CCD 8 408.
26 See Gruber 1981, p. 152.
27 Wallace (J) 1995, p. 5.
28 Wollaston 1860, p. 142.
29 Huxley 1893, vol. 2, p. 54.
30 Laugel 1868, pp. 136–137.
31 *Revue des deux mondes*, 1 March 1869, p. 89.
32 Royer in Darwin 1866 pp. 582–584 and Quatrefages' comments (1869, p. 90). For Royer, Darwin's "single prototype" must have been at least numerically multiple; there could not have been just a single, isolated individual.
33 Dally in Huxley 1868, pp. 25–26.
34 Valroger 1873, p. 160.
35 Broca 1870, p. 536.
36 On polygenism, see Blanckaert 1981 and 1996.
37 *Blackwood's Edinburgh Magazine*, v. 89 (544), February 1861 p. 165.
38 Owen 1866–1868, vol. 3, p. 814. See also Owen 1860a p. 514; 1860b, pp. 403–404: "Pouchet has contributed the most valuable evidence as to the fact and mode of the production by external influences of species of Protozoa."
39 Rossi 1870, p. 1.
40 *Ibid.*, pp. 65–66.
41 Puech 1873, p. 12.
42 See Darwin, "The doctrine of heterogeny and modification of species", *Athenaeum*, 25 April 1863, p. 554.
43 Valroger 1873, pp. 154–155.

44 Bastian 1872. See also Strick 1999 on the tensions between Bastian, supported by Alexander MacMillan, the editor of *Nature*, and the supporters of Darwin led by T.H. Huxley. Darwin himself was more favourable towards Bastian than was Huxley.

45 *Origin* 1859, p. 5 and 128, *Var* 74 (#37) and 272 (#394).

46 Bastian 1872, vol 2, p. 576. Here, Bastian is clearly leaning on Spencer.

47 *Ibid.*, p. 574.

48 Darwin 1868, vol. 2, p. 272.

49 Bastian 1872, vol. 2, p. 578.

9 "Light will be thrown"

Or on the origin of mankind

"One great difficulty to my mind in the way of your theory is the fact of the existence of Man". So wrote the naturalist and clergyman Leonard Jenyns (who happens also to be the brother-in-law of John Stevens Henslow) in a letter to Darwin.[1] Darwin made only the most incidental of references to the origin of man in the *Origin*, as though it were to be slipped in unnoticed. According to a tenacious interpretation, propped up by Darwin himself, all he does in the *Origin* is to indicate that, thanks to his theory of natural selection, "light will be thrown on the origin of man and his history".[2] But Darwin's account is contested and many have been increasingly vocal in asserting that "the origin of mankind" is, in actual fact, present throughout and weaved into the *Origin*, notably on the issue of sexual selection. This dispute can only arise because Darwin's readers paid little or no attention to Darwin's reserve. Just as we saw with the origin of prototypes, here again, in reflecting on the origin of mankind, it is another mere fragment of a sentence that has been seized upon: the promise that "light will be thrown" has been so often quoted and commented on that the very importance of the Darwinian theory itself seems almost to hang on the implications it has for the origin of man—or, as Jenyns abruptly phrased it, that "[man] is to be considered a modified and no doubt *greatly* improved orang!" "Light will be thrown": these few words from the *Origin* would go on to blacken a thousand pages. Darwin himself published *The Descent of Man* in 1871 and, even before that, many important publications had already begun defining the field of what would later become Darwinian anthropology.[3]

Humankind in the *Origin*

In truth, the question of "man" seems utterly absent from the *Origin*. The brief abstract Darwin sent to Harvard botanist Asa Gray on 5 September 1857, made no mention of sexual selection or human evolution. In the *Origin*, one finds the famous cryptic and prophetic passage which reads:

> In the distant future I see open fields for far more important researches. Psychology will be based on a new foundation, that of the necessary acquirement of each mental power and capacity by gradation. Light will be thrown on the origin of man and his history.[4]

Gillian Beer, for one, spoke of Darwin's "tactical" avoidance of the topic, and this represents the standard view on the question.[5] Darwin was personally convinced that selection, be it natural or sexual, had worked on humans, but he had resolved not to publish about it. Thus, at the very least, Darwin may have indulged in *alluding* to the topic but not *developing* it.[6] He had already settled his mind on this question when, working on his "big book", he stated to A.R. Wallace: "I think I shall avoid whole subject, as so surrounded with prejudices: though I fully admit that it is the highest and most interesting problem for the naturalist".[7] Darwin clearly had personal opinions regarding the issue. To Leonard Jenyns, he had written in January 1860:

> With respect to man, I am very far from wishing to obtrude my belief; but I thought it dishonest to quite conceal my opinion. Of course it is open to everyone to believe that man appeared by separate miracle, though I do not myself see the necessity or probability.[8]

However, this standard account has been challenged in different ways. First, Adrian Desmond and James Moore have claimed that, although mostly absent from the *Origin*, the human question had always been central to Darwin from the inception of his project. In their view, when Darwin writes in the *Origin*

> the doctrine of the origin of our several domestic races from several aboriginal stocks, has been carried to an absurd extreme by some authors. They believe that every race which breeds true, let the distinctive characters be ever so slight, has had its wild prototype,[9]

he *explicitly* refers to pigeons but *secretly* has humans on his mind, or so they argue. In fact, Darwin's initial manuscript devoted a certain number of passages to the question of human origins, but the topic was later dropped when Darwin wrote his "Abstract"—the *Origin*—thus occasionally drifting away "from its sacred cause".[10]

But another objection to the standard view was raised by Carl Bajema in the *Journal of the History of Biology* in 1988. The famous passage on "light will be thrown", he claimed, did not mean that Darwin would not refer to man, thus sparking a debate among Darwin scholars.[11] Humans, Bajema showed, are actually present in the *Origin*: Darwin often refers to "uncivilised man", to "savages", to "savage races of man", even to what a "savage" may have known or may think today when looking at a ship.[12] Here, it may be wise to distinguish between references to *human cultural evolution* and what would be *an actual evolutionary treatment* of the human species and its origins.[13]

However, it could be added that, in another passage of the *Origin*, Darwin mentioned specifically that "some little light can apparently be thrown on the origin" of "the differences between the races of man, which are so strongly marked [...] chiefly through sexual selection of a particular kind, but without here entering on copious details my reasoning would appear frivolous".[14] Those considerations on humankind into the *Origin* are of great importance, but of limited scope and size. It is quite clear that, as Peter Bowler noted, "by the time Darwin published the *Origin of Species* in 1859, no one could be in any doubt as to the implications of applying the theory of evolution to mankind".[15] But it is also striking that the "origin of mankind" question never became prominent within the *Origin*. As we saw in our first chapter, the *Origin* changed a lot and grew by about one third from 1859 to 1872, but it was by no means ever intended to absorb Darwin's opinions on every possible topic, even though this may have been eagerly wished for by his readers.[16]

To focus the origin question on humankind is to sell short the originality of Darwin's method. Where the naturalists of old were scholars whose only preoccupation was humankind, Darwin sparked a revolution by placing animals at the fore.[17] The question of origin, first formulated with languages and civilisations in mind, had now been extended to include biological species. In a way, we backpedal towards anthropocentrism by demanding that Darwin provide the origin of man, as if the origin of species should not be decided generally but only in strict accordance with the specific issues implied by our own species.

From the *Origin* to the *Descent*

At the time he wrote the *Origin*, Darwin already knew how to deal with the question of humans (by means of sexual selection), but he never gave full development to his views on this question in the *Origin*.[18] In fact, it seems that Darwin never even planned to devote an entire book to "man"; rather it came about as an outgrowth of his other works. In a letter to Alphonse de Candolle, written soon after the publication of his book on *Variation*, Darwin stated:

> I have had the manuscript for another volume almost ready during several years, but I was so much fatigued by my last book that I determined to amuse myself by publishing a short essay on the *"Descent of man"* ... Now this essay has branched out into some collateral subjects, and I suppose will take me more than a year to complete.[19]

On the road from *Origin* to *Descent*, one is struck by the sharp contrast between the two books. For instance, on the issue of gender differences between men and women, Darwin presented his ideas in the *Descent* in quite a different tone from the neutral tack adopted in the *Origin*.[20]

Cross-referencing between the two texts, *Descent* and *Origin*, deepens rather than resolves the problem of Darwin's own doctrinal coherence. Yvette Conry, in particular, has identified in the divide between the two books: "the shift, in the text itself, from Darwin to Darwinism, from a scientific theory to an ideological investment". She investigates the phenomenon of "extra-scientific hijacking" and the way in which "a myth comes to replace a science".[21] Conry's interpretation enables us to highlight absolutely decisive conceptual inflections. In particular, she makes visible a clear epistemological reorientation of natural selection; a simple "principle" in the *Origin*, tasked with summing up and coordinating the facts of individual variation and the geometrical rates of population growth, by the *Descent* it has become an "inflexible law" functioning as a "law of progress". In another important inflection, the positivity of artificial selection is seen to increase from one text to the next; its analogical and instructional role within the *Origin* is injected to take on a reinforced function in the *Descent*, whereby "civilised" is identified with "domesticated". Conry's results are enlightening. One could, nevertheless, choose to hold fast to the separation between an ideal or perfect "Darwinism" and Darwin the man, historically contingent and necessarily imperfect. By giving up on readings of Darwin that pick out only what corresponds to the ideal of "Darwinism", by agreeing to read him in a manner respectful of the actual words he communicated to us, the question would certainly take a different turn.

Conry describes the relation between the *Descent* and the *Origin* as one of ideological application and epistemological decline. However, with the same interpretative logic, other conclusions can be reached. Where Conry sees methodological reorientation and ideological drift, others, such as Clémence Royer, draw quite the opposite conclusions. The same interpretative postulates—the decision to extract the logic from the system rather than reading what Darwin wrote—saw Royer rejoicing at the publication of the *Descent*. Whereas Conry endeavours to pull pure "Darwinism" away from its envelope of social interpretation, Royer rushed avidly to exact its social lessons. And yet both seemed to suggest that even Darwin had not understood himself.

Royer's Preface to the *Origin* emphasised the moral impact she assumed Darwin's theory had for humans. As she quite infamously wrote:

> Data from the theory of natural election can no longer allow us to doubt that the superior races have been produced successively, and that therefore, by virtue of the law of progress, they have to be destined to supplant the inferior races by their further progress—instead of miscegenation and confusion, which put them at the risk of becoming absorbed into them through interbreeding which lowers the average level of the species.... Men are different by nature: that is the point from whence we must start.[22]

Royer shows no hesitation in considering that "the oh-so controversial problem of human origin could be seen as Darwinism's real stumbling block".[23] Royer's "stumbling block" is ambiguous: it is what the truth of the theory comes up against, the point at which it shall stand or fall, fated to emerge as either established and true, or false and rejected. Darwinism's logical coherency is sharpened by its application to the central anthropological problem: how does Darwinism explain not just the origin *of the* species, but the origin of *that particular* species that is human? Although Darwin somewhat conspicuously underlines the fact that for the longest time he had made a point of not broaching this problem, certain readers have not hesitated in identifying it as the "great battlefield of Darwinism".

Polygenism and monogenism

At first, the problem of the "origin of mankind" refers to the origin of a hypothetical ancestry common to men and the anthropoid apes, as well as the origin of our relation to the gorilla, the chimpanzee, and the orangutan. Darwin was well aware of debates concerning human races, especially in reaction to Richard Owen's stance on the question.[24] After Owen had given his lecture, "On the Characters, Principles of Division, and Primary Groups of the Class Mammalia", Darwin wrote to Hooker: "Owen's is a grand Paper; but I cannot swallow Man making a division as distinct from a Chimpanzee, as an ornithorhynchus from a Horse: I wonder what a Chimpanzee wd. say to this?"[25]

But the problem of "origin of mankind" also applies the scheme of common descent to the various races of man within the human species. The case of Harvard ichthyologist Louis Agassiz (1807–1873) sheds interesting light on the issue of the origin of species in relation to the common descent of mankind. For Agassiz, speaking of the "origin of species" is an oxymoron: "If species do not exist at all, as the supporters of the transmutation theory maintain, how can they vary? and if individuals alone exist, how can the differences which may be observed among them prove the variability of species?"[26] Ernst Mayr linked Agassiz's rejection of Darwin's ideas to his commitment to idealistic philosophy and his concept of species as a Platonic *eidos*.[27] But a revisionist account by Mary P. Winsor showed that Agassiz's view of species "was more interesting and complex than the label 'typologist' suggests". Winsor emphasised both Agassiz's epistemological standards, which the idea of evolution did not meet ("the narrow road of fact-based inference"), and his embroilment in the debate on human races in America. Once in the United States, Swiss-born Agassiz realised that "to insist that mankind was nevertheless descended from one common ancestor would be to demonstrate evolution".[28] Within the debate on human races, the criterion of distinct species could not be the ability to interbreed, but something like "primordial organic form", as race-theorist Samuel George Morton put it—a definition Agassiz whole-heartedly applauded.

The question of the human species' origin, framed within anthropology, takes on a particular resonance that partially covers over or defines a new form of the opposition between "polygenism" and "monogenism"—the same question we have considered regarding the origins of life. On the one hand, there is the hypothesis that the different races resulted from distinct creations; on the other, the hypothesis affirming that all men share the same origin. Just as in the case of the unity or multiplicity of seeds, the application of Darwin's views to anthropology brings about a shift of positions. Where anthropological monogenism might easily have appeared to be a relic of creationism, Darwin's views reinvigorated it, opening the way to positing adaptive mechanisms behind the diversification of humankind; conversely, polygenism, the old libertine weapon of choice in combating the biblical account, now comes to be seen as a suspect theory, opening the way to separate creations. Of the two terms representing the two sides of this alternative, which one is best supported by the Darwinian structure?

What actually occurred was a shift in the controversial value of polygenism: "It approaches comedy!" a contemporary commented.

> Before Darwin's book appeared, defenders of a primitive unity of human races were considered to be backward *relics*, removed from all scientific progress. Now, it is held to be beyond doubt that apes and men share a common ancestor of an intermediary form.[29]

As for Huxley, he attributes Darwin with the discovery of a way

> of reconciling and combining all that is good in the Monogenistic and the Polygenistic schools [...] It is true that Mr. Darwin has not, in so many words, applied his views to ethnology; but even he who "runs and reads" the *Origin of Species* can hardly fail to do so; and, furthermore, Mr. Wallace and M. Pouchet have recently treated of ethnological questions from this point of view.[30]

In other words, Darwin can still teach us how to respond to problems even if he himself did not take them on.

To apply, or not to apply, to humankind

Anthropology might be used to transform Darwin's views into a local doctrine and take away its status of generality. According to James Hunt, President of the Anthropological Society of London in 1864–1865, "Darwinism may be true when applied to botany or zoology; but there is not a fact in the whole range of anthropology which lends it any support".[31] This indicates the creation, throughout the field of anthropology, of a tension with respect to the coherency of "Darwinism" and its capacity to become a general theory.

As Auguste Laugel explained in 1868, the human mind is so drawn to Darwinian theory because it recognises that, if the theory be true, then it must apply to man as well as animals. This is precisely what drove enthusiastic disciples and fervent critics alike to overstep Darwin's own "prudent reserve" and "enigmatic silence": they "pushed the system to its very last logical consequences and, upon their faith, a great number of souls believe these consequences to be damaging to our species, detrimental to our greatness and our dignity".[32]

That the origin of mankind be so fundamental is because the very foundations of morality and the nature of human intelligence rest upon it. Some wish to deny "Darwinism" in order to save morality, while others envision the sacrifice of morality upon the altar of Darwinian fervour. Those who attempt to dissociate the two problems are few and far between. Huxley, however, asked: "Is mother-love vile because a hen shows it, or fidelity base because dogs possess it?" Émile Ferrière would have chorused, "No, a thousand times no! Virtues are noble in and of themselves, regardless of the form surrounding them. *Morality stands independent of all social conditions and all origins*".[33] He implores obedience to the inescapable, imperious duty to reason, to the brotherly call of the heart, what matter if such instincts be those of a perfected ape!

The strange case of A.R. Wallace

Independently of these conjectural moral consequences, what real logical consequence results from the hypotheses advanced by Darwin? The case of Alfred Russel Wallace provides a perfect illustration on this point. The relationship between Wallace and Darwin had always been that of a brilliant right-hand man to his superior, who recognised not only Darwin's priority but also his foresight. Wallace long embodied the radical and absolute selectionist, firmly set on seeing natural selection at work everywhere, underpinning everything from the beauty of the peacock's ornamentation to the mimetic disguise of the butterfly's wings. The very same Wallace, however, quite surprisingly placed mankind outside the dominion of natural selection. Of all the differences between Darwin and Wallace, this is the principal (if not the only) one to have been widely identified by their contemporaries. The origin-of-man problem is an entry point for pitting Wallace (inconsistent Darwinian) against Darwin (consistent, but sometimes silent).[34] Some, such as Edouard Claparède in Switzerland, or Theophilus Heale of the Auckland Institute in New Zealand, place Wallace among the "opponents or rather reformers" of Darwinism: far from seeing in Wallace a Darwinian, they set him alongside those who limited the power of natural selection.[35] The gap between Wallacism and Darwinism, already emphasised by Romanes (see Table 6.1), is now widened as the origin of species is reformulated into the origin (and maintenance) of the superior intellectual faculties. It is Wallace himself who defines the dimensions of this gap. In a chapter of his *Contributions to the theory of natural selection* entitled "The limits of natural selection as applied to man", he explains. "What Natural Selection can Not do".[36] Natural selection bears on the principle

> that all changes of form or structure, all increase in the size of an organ or in its complexity, all greater specialization or physiological division of labour, can only be brought about, in as much as it is for the good of the being so modified.

Hence, it

> has no power to produce absolute perfection but only relative perfection, no power to advance any being much beyond his fellow beings, but only just so much beyond them as to enable it to survive them in the struggle for existence.

But, above all, Wallace concludes that natural selection cannot produce variations which would be in any sense injurious to the individual "on their first appearance"; such variations "could not possibly have been produced by natural selection. Neither could any specially developed organ have been so produced if it had been merely useless to him, or if its use were not proportionate to its degree of develop-ment".[37] Such cases as these would prove that some other law, or some other power, then natural selection had been at work. Now if we consider what changes were essential to the full moral and intellectual development of human nature,

> we should then infer the action of mind, foreseeing the future and preparing for it, just as surely as we do, when we see the breeder set himself to work with the determination to produce a definite improvement in some cultivated plant or domestic animal.

Wallace asserts that inquiry into what natural selection can or cannot do, is "as thoroughly scientific and legitimate as that into the origin of species itself": "It is an attempt to solve the inverse problem, to deduce the existence of a new power of a definite character, in order to account for facts which according to the theory of natural selection ought not to happen". In other words, pursuing the Darwinian master stroke which consisted of placing origin positively within the scientific field, Wallace proposes another course of action; to seek the cause of those facts which escape theory. He guarantees us that this cause lies "strictly within the bounds of scientific investigation".

The gap between Darwin and Wallace regarding man's place within the Dar-winian structure constitutes a particularly slippery pitfall: it is singled out by critics as a dissension in the Darwinian school, a source of potential "schisms", and commented upon with great surprise, as some bizarre quirk, by Darwin's followers.[38] Wallace's supposed "inconsistency" irritated believers in Darwin-ism's coherency yet didn't convince opponents of it: Wallace was to be con-demned on all fronts. He made the mistake of not respecting Darwinism's demand for coherency and found himself faced with a dilemma: if a higher power had intervened in the formation of human races, then why not also look to that power to explain the creation of all other animal and plant species? If, on the other hand, natural selection alone constitutes the well-founded explanation for this latter creation, then is it not consequently *un*founded to call upon some higher power in order to account for the formation of the human races?

Wallace protested against and opposed a narrow conception of Darwinism whose aim was to squeeze it into a logical box. His protestations came to naught. He was called upon by the field of anthropology to shed light on the origin of man "in the name of the theory he had helped found". However, in every answer, he preferred to "confess the doctrine's impotence in treating of questions of the specific attributes of the human species".[39]

Further away: driving Darwin's theory to its ultimate consequences

Darwin's doctrinal coherency forms a programme that is shared by several of his successors. In an article published as "A deduction from Darwin's theory", William Stanley Jevons claims the following:

There is one important consequence deducible from Darwin's profound theory which has not yet been noticed so far as I am aware. The theory is capable under certain reasonable conditions of accounting for the fact that the highest forms of civilisation have appeared in temperate climates.[40]

Stating that "it is no doubt true that man displays his utmost vigour and perfection, both of mind and body, in the regions intermediate between extreme heat and extreme cold", Jevons claims to provide a serious explanation, in contrast to "the explanations hitherto given of this fact [which] are of a purely hypothetical and shallow character". In order to do this, Jevons' intention was to shift from principles to their consequences, excessively and improperly using the vocabulary of logical consequence throughout his text: "It will of necessity follow". The theory of natural selection represents "that great method by which infinitely numerous adaptations will always be produced throughout time". For Jevons,

> the essential consequence of Darwin's views [is] that no form of life is to be regarded as a fixed form; but that all living beings, including man, are in a continual process of adjustment to the conditions in which they live.

So, "it will of necessity follow that the longer any race dwells in given circumstances, the more perfectly will it become adapted to those circumstances". In this way, Jevons believes he can deduce a relation between sedentism and adaptation to environment. Framed in this way, migrant peoples are disadvantaged, and Europe, which Jevons considers to be the oldest settled part of the world, must also be home to the most perfect humans. He concludes

> that the utmost result of speculations of this kind, supposing them to be valid, would consist in establishing a *general tendency*, so that the probabilities will be in favour of a great display of civilisation occurring in temperate climates rather than elsewhere.

The coherency of Darwin's views is also at the heart of *Le Darwinisme*, a brief opuscule written by Émile Ferrière (1872), which questions the solidarity between a theory and its applications. He posits that some theories (doomed to rapid decline) attempt to construct a system and then bend facts to it; other theories involve only the bonds relating "real and constant" facts to each other. Abuse or misuse may still be made of them, but they constitute a "method of research" and evolve within a bounded circle wherein they are certain and solid. A good theory, therefore, is efficient when placed in "prudent hands" which will bring it to "services which, while perhaps slow, are serious and continuous".

This criterium of prudence allowed Ferrière to distinguish between good and bad applications of a theory, prune any mishandling, slice away any excess, without, however, attacking the core of the doctrine. How, then, does this relate to the theory of evolution? Some naturalists bring it to bear on the origin of species mystery; others propose to employ it in a more radical way, in an "attempt to relate all genera and all species back to just three or four primordial types". Importantly, it is possible for the latter enterprise to fail without this having any repercussions on the former: "If it can be applied to species only, reducing their artificial multitude to a small but real number, then this will suffice for it to conserve great value in natural history". This approach would, for instance, allow for Herbert Spencer (who applies the theory of evolution to sociology, politics, morality, etc.) to be either embraced or rejected without either decision bringing the naturalist part of the theory under attack. It also allows the putative status of linguistics (as the theory of evolution's principal field of effectivity) to be determined:

> with no other science, not even natural history, does the theory of evolution fit so well as with the history of languages. The application is of such striking precision, it could truly be thought that the theory of evolution had been born of philology.[41]

Thus, Ferrière's text tasks itself with presenting the principles of Darwin's theory before then applying them to the evolution of languages and the evolution of humankind, "the great battlefield of Darwinism".[42]

For many, Darwin's theory of descent must just be traceable to the *Ursprung* rather than being only a description of the *Entstehung* process. Understood in this way, "Darwinism" seems to be leading to the infamous "ape-man" theory, to the idea that man's ancestry is simian. It is not only by interpretation of the word "origin" (understood as the search for a genealogical tree) that the origin of man imposes itself on Darwinian ground. The argument of a necessary correlation between Darwin's views and the simian origin of humans, although the dominant view, is not the only one possible, and some have maintained that "Darwinism" does not necessarily lead to the ape-man vision. This is the case with Wallace, who sets man's superior faculties apart from and above the rule of natural selection. It is also the case with Quatrefages, for whom, by reasoning logically from the principles of Darwinism, there is a contradiction in having man descend from apes. We see from this that the problem of applying Darwinian theory to man is framed notably by Haeckel's genealogical interpretation and the ambiguity of the Wallacian position. Although several radical "Darwinians" (Royer, Haeckel) do seem to agree with their enemies (*Civiltà cattolica*, Lecomte) in affirming that Darwin's views necessarily result in the ape-man theory, several other equally fierce "Darwinians" (Wallace, Quatrefages) conclude that the simian origin theory of man sits in plain disagreement with Darwin's ideas and that only erroneously could anyone try to force them together.

Having opened the door to unifying diverse sets of phenomena, Darwin cuts a bold figure: the man who didn't hesitate to relate all vertebrates to a single common root. Regarding the human species, the silence Darwin scrupulously maintained throughout the *Origin* is an eloquent mark of his clairvoyance. His saying nothing must have been to (provisionally) conceal consequences which would only have scandalised his time. Yet everything was poised and in place: if

Table 9.1 Darwinism applied to species and languages (according to Ferrière 1872)

Selection:

among species	among languages
1 Species have their varieties, produced by the milieu or physiological causes	1 Languages have their dialects, produced by the milieu or the habits
2 Living species generally descend from other species from the same country	2 Living languages generally descend from other dead languages from the same country
3 In an isolated country, a species produces less variations	3 In an isolated country, a language produces less variations
4 Variations induced by crossing with distinct or foreign species	4 Variations induced by introduction of new words, due to foreign relations, science and industry
5 Superior physical qualities assure victory to the individuals of one species: a cause of selection	5 Literary genius and centralised instruction: a cause of selection.
6 Beauty in plumage or singing: a cause of selection	6 Brevity or euphony: a cause of selection
7 Numerous gaps in extinct species	7 Numerous gaps in extinct languages
8 The duration of a species is dependent on the number of the individuals who compose it	8 The duration of a species is dependent on the number of individuals who speak it
9 Extinct species will never reappear	9 Extinct languages will never reappear
10 Progress in species by division of physiological labour	10 Progress in languages by division of intellectual labour

Genealogical classification:

among species	among languages
1 Structure remaining unchanged; organs of high physiological importance; organs of varied importance	1 Structure remaining unchanged; radicals of high importance; flexions of varied importance
2 Vestiges of primordial structure: atrophied or rudimentary organs, embryonic structure	2 Vestiges of primordial structure: atrophied or rudimentary letters, embryonic sentences
3 Uniformity of a set of characters	3 Uniformity of a set of characters
4 Chain of affinities in the living or extinct species	4 Chain of affinities in the living or extinct languages

"Darwinism" is a process of generalisation, then it must be carried to its conclusion and extended even further.

Notes

1 Jenyns to Darwin, 4 January 1860, CCD 8 14.
2 *Origin* 1859, p. 488, *Var* 757 (#257).
3 Cf. Vogt 1863; Huxley 1863; Haeckel 1866; Haeckel 1873. See also Bowler 1986, Richards (RJ) 1987 and Radick 2013.
4 *Origin* 1859, p. 488.

5 Beer 1983, pp. 58–59. See also, Mayr 1982, p. 438.
6 Cooke 1990.
7 Darwin to Wallace, 22 December 1857, CCD 6 515.
8 Darwin to Jenyns, 7 January 1860, CCD 8 25.
9 *Origin* 1859, p. 19.
10 See Desmond and Moore 2009, especially Chapter 11, p. 311: "Man was untouchable in the *Origin*, the human races are too sensitive…".
11 Bajema 1988.
12 *Origin* 1859, resp. p. 38, 140; 18, 34, 36, 38, 42, 198, 215; p. 382; p. 17, 485.
13 Bowler 1989b.
14 *Origin* 1859, p. 199.
15 Bowler 1986, p. 2.
16 Bizzo 1992.
17 Virchow, 1882, p. 418.
18 Richards (E) 2017.
19 Darwin to Alphonse de Candolle, 6 July 1868.
20 For a more radical (and critical) account, see Erskine 1995.
21 Conry 1987, pp. 63–64, 67, 102.
22 Royer 1862, p. lxi.
23 Royer 1866, p. iii.
24 Rupke 2009, Chapters 6 and 7. Delisle 2016, p. 48 (on Owen).
25 Darwin to Hooker, 5 July 1857, CCD 6 419. See Owen 1858.
26 Agassiz 1860, p. 143.
27 Mayr 1976, pp. 251–276.
28 Winsor 1979, p. 92.
29 Rudolph Wagner, quoted in Valroger 1873, p. 68.
30 Huxley 1893, vol. 7, p. 248.
31 Quoted by Dally 1868, p. 36; Lecomte 1873, p. 39.
32 Laugel 1868, p. 130.
33 Huxley 1863, p. 111; Ferrière 1872, p. 177, emphasis by the author.
34 On Wallace's views on man, see Kottler 1974 and Smith 1999.
35 Claparède 1870, p. 565; Heale 1872, p. 446.
36 Wallace 1870, p. 333.
37 *Ibid.*, p. 334.
38 See, for example, Valroger 1873, pp. 107–108, Royer 1880, pp. 765–766.
39 Quatrefages 1877, p. 84.
40 Jevons 1869, p. 231.
41 Ferrière 1872, p. 6.
42 Ferrière 1872. On Darwinism and the origin of language, see Radick 2007, Chapter 1.

10 Darwin-the-Darwinist, or the quest for systematic coherency

When we speak of the quarrels or controversies surrounding "Darwinism", what often comes to mind is the violent opposition between the protectors of religion and the defenders of evolution. This perspective traces the outlines of certain socially contrasting schemas; between conservative aristocracies and radical plebeian movements, amateur scientist theologians and professional savants steeped in atheism.[1] But, strikingly enough, both sides claim to have discovered the "maximum logical coherency" of Darwin's views: they too look for a certain logical coherency of "Darwinism", sometimes even in contravention of Darwin's own words.

The search for the logical coherency of "Darwinism" pushed enquiring minds to overstep the actual word of the *Origin* in order to find out where the system itself actually led to, without having ever explicitly stated this endpoint. What we have uncovered are two distinct frameworks into which the *Origin* has been pulled and into which Darwin's question (the origin of species) has been shifted. In one, the origin of life and of prototypes; in the other, the origin of man, of superior intellectual faculties, and of societies. In each of these camps, various positions set themselves the task of formulating what is the genuinely Darwinian theory (specifically, an opposition between mono- and polygenism). Each of these positions is, in turn, open to various readings. Does it lead to heresy and atheism? Or is it compatible with orthodoxy and the sacred texts? Every one of these readings shifts the central core of Darwin's texts to some radical consequences. Consequently, it empties the *Origin* of its substance, reinterpreting it through the pinhole lens of mere sentence fragments torn from its final pages.

I call these attempts "radical readings of Darwin": "radical readings" force some unexpected consistency into Darwin's own words, ideas, texts, often going against the grain of Darwin's acknowledged and published claims. They turn Darwin into a Darwinist. We have seen just such radical readings in the previous chapters, on the issues of life and humankind. Radical readings can support views that supporters of "Darwin-the-Selectionist" would identify as authentically Darwinian; but other radical readings can produce views that panselectionists would call monstrously "pseudo-Darwinian", even "Lamarckian" views. Clémence Royer, for instance, is just such a radical reader of Darwin when she develops her views on what would today be called "Social Darwinism".

The search for "radical coherence" is one way of reading Darwin, just like "limited relevance" might be another. A.R. Wallace, although a radical selectionist for non-human species, considered the human mental powers to lie beyond the reach of natural selection and, therefore, strongly opposed supporters of "radical coherence". The "radical coherence" strategy can be claimed by both supporters of Darwin (they say they are merely extending his views) and by adversaries (who try only to identify monstrous consequences of Darwin's views). Another point for consideration is that Darwin was not himself a "radical", in the sense that he did not look for "maximum logical coherence": his strategy in the *Origin of Species* was clearly to de-emphasise any radical consequence some may have wished to draw from his book. Typically, he deemed as irrelevant such issues as the origin of human psychological powers or the origin of life.[2] These he carefully avoided in the *Origin of Species* (although they are broached in the notebooks and private letters).

The aim of this chapter is to examine the relation between the *Origin* and certain religious, social, and political controversies through the optic of the doctrinal coherency or supposed unity of the general system of "Darwinism". Read from this perspective, it appears that the unity of "Darwinism", taken as a coherent vision of the world, possesses a certain "amphiboly" in the sense that the coherency of Darwin's *Origin* has implications that point in two different directions. On the one hand, unified Darwinism is an atheist system, a reductionist naturalism; but, on the other, its unity acts as a sign of harmony between our intellectual faculties and the structures of the world. When one considers both sides of the problem, the coexistence of the two arguments translates an agreement between theory and reality, which, in turn, acts as a guarantee of the divine rationality present within nature itself. This Darwinian amphiboly—where the unity of Darwinism is cast either as atheist henchman or as divine guarantee—explains how the system's unity and coherency can be just as easily mobilised by the materialist school (who call upon it as a healthy cure against superstition) as by the spiritualist school (who read in it the traces of Providence).

For its radical readers, "Darwinism" cannot be a local theory, it has to become a general theory since Darwin's question itself demands that its broader implications be sought out. First, it was Darwin himself who began the trend by pushing his arguments to their final term, referring to the *origin* of species, when he could just as well (in keeping with the opinion found amongst his contemporaries) have considered the *diversification* of species into (intra-species) varieties. In a nutshell, the movement towards the generalisation and systematisation of the Darwinian theory into a Darwinist system is there, in embryonic form, in the very title and goal of Darwin's book. Still, many readers resisted. Carl Nägeli, for one, subtly changed Darwin's question of *Entstehung* (origin) into one of mere *Veränderung* (modification).[3] Asa Gray also noted a similar problem when speaking of the final pages of the *Origin*:

> Why should a theory which may plausibly enough account for the diversification of the species of each special type or genus be expanded into a

general system for the origination or successive diversification of all species, and all special types or forms, from four or five remote primordial forms, or perhaps from one?[4]

The very project of the *Origin of Species*, as formulated in its title and presented again in its closing pages, calls for a radical and coherent approach.

Moreover, "Darwinism" makes itself an integral part of the search for that unique law which would rule over both the organic and the inorganic; so, in this too, it presents itself as a system that does not suffer limits and that demands a global approach. Such radicalism was already apparent in Robert Chambers' *Vestiges*, wherein the author underlined the reach held by the rule of law in the universe: "the whole appears complete on one principle".[5] For Chambers, law formed space, law made it into "theatres of existence for plants and animals"; then law developed "sensation, disposition, intellect". For Chambers, two great comprehensive laws, gravitation and development, rule over two departments, the inorganic and the organic; both fields may even be "only branches of one still more comprehensive law, the expression of a unity, flowing immediately from the One who is First and Last". Hence the unity between the domains of being is reflected in the progress of science towards the unity of a comprehensive law, the very expression of divinity. Scholars are invited, not to touch the essence of the organising force, but to ask themselves what it might actually be, to apprehend each of its manifestations and then to coordinate them.

Such theories of development are comparable to the theory of conics, which sequences four different curves (circle, ellipse, parabola, hyperbola) of quite diverse, if not opposing, properties and natures. At first the curves appear to be distinct, but, as Herbert Spencer first observed in 1852, to go from one form to another, it is really just a question of appealing to "a single process of insensible modification".[6] Likewise, the reduction of all physical forces (calorific, electrical, magnetic) to a single force indicates that phenomena from diverse levels can indeed result from one identical principle, of which they would be only the transformation.

This quest after the laws of nature runs through the *Origin*, from beginning to end. It is the very meaning underlying the quote from Whewell inscribed by Darwin on the first page of the book:

> with regard to the material world, we can at least go so far as this—we can perceive that events are brought about not by insulated interpositions of Divine power, exerted in each particular case, but by the establishment of general laws.

The same quest also closes the book, in the "entangled bank",[7]

> clothed with many plants of many kinds, with birds singing on the bushes, with various insects flitting about, and with worms crawling through the damp earth, and to reflect that these elaborately constructed forms, so

different from each other, and dependent on each other in so complex a manner, *have all been produced by laws acting around us.*

Like Chambers, Darwin explains "this view of life, with its several powers, having been originally breathed into a few forms or into one" using a parallel between the view of "this planet ... cycling on according to the fixed law of gravity" and the fact that "from so simple a beginning endless forms most beautiful and most wonderful have been, and are being, evolved".[8] Where Chambers had drawn a parallel between gravitation and development, Darwin does the same with gravity and evolution.

Darwin's manifest insistence on laws earned him a reputation as a "Kantian": in substance, he recognises no other forces outside those of physics and chemistry, even if he does allow for biological phenomena which transcend the physico-chemical.[9] In this case, does Darwin ultimately do nothing more than substitute several new laws, the law of natural selection being one of them, for Chambers' and Spencer's "development"? Might this be his only mark of originality?[10]

Because its strength is in its proposed unification of various fields of biology, Darwinism could never limit itself to being a local theory. Its scope must encompass the whole (living) world. Much less is it possible to reject the transmutation of species from a single prototype under pretext that it would lead to atheism. In fact, Asa Gray advanced that it was the opposite argument which must be upheld: because Darwinism unifies, it cannot be atheist. The value of such objections to the theory of derivation can be tested:

> The common scientific as well as popular belief is that of the original, independent creation of oxygen, and hydrogen, iron, gold, and the like. Is the speculative opinion now increasingly held, that some or all of the supposed elementary bodies are derivative or compound, developed from some preceding forms of matter, irreligious? Were the old alchemists atheists as well as dreamers in their attempts to transmute earth into gold? Or, to take an instance from force (power)—which stands one step nearer to efficient cause than form—was the attempt to prove that heat, light, electricity, magnetism, and even mechanical power, are variations or transmutations of one force, atheistical in its tendency? The supposed establishment of this view is reckoned as one of the greatest scientific triumphs of this century.[11]

Darwin unifies biology through derivation, just as others unified mathematics, chemistry, and physics before him: there is nothing atheistic about it.

However, numerous readers of Darwin did not lend him such an open ear. Materialism and atheism are associated with the transmutation of species via the dual opposition of final causes/Epicureanism and utility/chance. Similar rhetoric is employed by Darwin's opponents, such as Paul Janet,[12] and followers, like Carl Vogt, for whom "it is indeed indubitable that Darwin's theory

unceremoniously dismisses the personal creator, with its alternative intervention into the transformations of creation and the apparition of the species; it
leaves absolutely no place for the action of such a creator".[13] Law takes the
place of the "personal creator", to such an extent that, once the starting point
and the first organism have been provided, "all of creation develops in a continuous manner through natural selection, in accordance with the simple laws
of hereditary transmission". In the Darwinian world, mechanism is master.
Once the first elements are present, all the rest (including humankind) unfolds
in a regulated manner, without supernatural intervention: "as soon as the theory's
necessary conclusion had been understood, the storm broke out over everything, and it still has not passed".[14]

Law-based development takes over from the miracle of special creations.
While this does contradict the Revealed word, it is not necessarily atheistic. It
fits perfectly with a certain theism: not a personal God, but the unique substance
God of Spinoza. Many readers used this angle to turn Darwin into a reformer of
natural theology.

As a result, the search for Darwinism or Darwin's doctrinal coherency
answers to two different logics, what I call its "amphiboly" (see Table 10.1).

The *Origin*, as a system, is refuted either through analysis of the facts that it
presents, or through exposure of its required consequences. If the system is
attacked using the very facts it organises, then the cohesion of the presented
propositions can be challenged, a list of the theses it defends could be drawn up
and analysed for compatibility; its potential implications can also be sought out,
the system's untraced pieces can be drawn and the problem of missing links
exposed. One can then read the *Origin* through the lens of these consequences,
going beyond what Darwin was able to explicitly mention in his work. The
evaluation of Darwinism one reaches may then be either logical (what coherency
is there in the ideas presented?) or moral (are these consequences acceptable or
not?). For those who read Darwinism as a coherent system, the actual figure of
Darwin tends to retreat and the *Origin* is read as a simple contribution to the
grand doctrine of evolution.

In 1868, Huxley affirmed that

> perhaps this doctrine of Evolution is not maintained consciously and in its
> logical integrity, by a very great number of persons. But many hold par-
> ticular applications of it without committing themselves to the whole; and
> many, on the other hand, favour the general doctrine without giving an abso-
> lute assent to its particular applications.

In fact, Huxley believes,

> the only complete and systematic statement of the doctrine with which I am
> acquainted is that contained in Mr. Herbert Spencer's *System of Philosophy*,
> a work which should be carefully studied by all who desire to know whither
> scientific thought is tending.[15]

Table 10.1 Two readings of Darwin's logical consistency

Unity of Darwinism		Its doctrinal coherency, as perceived by:	
		defenders of the Origin	critics of the Origin
Coherence of the Darwinian system with the facts. Does the *Origin* account for available scientific data?	On the epistemological level	Darwinism accounts for the results of artificial selection; for palaeontological discoveries which confirm predicted results; for the structure of the classification	The Darwinian system is a simple hypothesis; it has no factual support, is even contrary to the facts, contrary to the characteristics of variation (limits, leaps), to the physical data about the age of the earth, and to the existence of species
			The lack of palaeontological data and missing links is bemoaned
	On the religious level	The scriptures must be interpreted in light of scientific discoveries (following the Copernican precedent)	The system is an imagining which contradicts scripture: it supposes infinite time, jeopardises morality, degrades man
		The Darwinian mechanism opposes the theory of special creations but does not exclude the intervention of a Creator. It is compatible with a theist interpretation.	It is a naturalist mechanism which detracts from the Creator but, ultimately, cannot do without him (Pascal's "chiquenaude")
Coherence of the Darwinian system with its supposed consequences	On the logical level	Science can work on laws without providing first terms, can give the *Entstehung* and leave aside the *Ursprung*	The system leads to untenable positions. It does not provide what it promises (the origin)
			It hangs suspended in a void
Beyond the word of the text itself, where does the *Origin* lead us?	On the moral level	Human morality is not degraded by being shared by apes and dogs	The system leads to unacceptable position. It is used to degrade man by having him descend from apes

Huxley depicts various ways of being committed to the theory of evolution:

> one who adopts the nebular hypothesis in Astronomy, or is a Uniformitarian
> in Geology, or a Darwinian in Biology, is, so far, an adherent of the doctrine
> of Evolution. And, as I can testify from personal experience, it is possible to
> have a complete faith in the general doctrine of Evolution and yet to hesitate
> in accepting the Nebular, or the Uniformitarian, or the Darwinian hypo-
> theses in all their integrity and fullness. For many of the objections which
> are brought against these various hypotheses affect them only, and even if
> they be valid, leave the general doctrine of Evolution untouched.[16]

These reservations about Darwinian theory and its place in the broader frame
of an evolutionist doctrine are liable to lead to certain shifts. The origin of
species is recast in the following declination: starting from the first organism,
once this primitive point has been provided, the object is to see how everything
else follows on; or, from principles, the investigation involves the consequences
of the doctrine on the status of humans or on the origin of living forms taken as
prototypes. The coherency of Darwinism proves itself further by its predictive
value, as well as by its ability to serve as a model for uniting more and more
fields. In the same way that Newtonianism saw a blossoming of local or specific
laws of attraction, the various "Darwinisms" resulted in a multiplication of novel
selections and struggles for existence.

Does the doctrinal coherency of Darwinism have to be established? This
question can be understood in several ways: how does it account for the facts it
encompasses? To which facts could it be legitimately extended? Are its explanatory principles shared by various sciences or are they, on the contrary, specific
to each one separately?

In short, the coherency can be understood either forwards or backwards, both
in actuality and potentiality: what Darwinism presents *in fact*, and what it could
do *if it were pushed*. In other words, there is a Darwinism-in-fact (found in the
pluralist maze of Darwin's works) and a Darwinism-in-principle (extrapolated
from Darwin's theories or statements). If Darwinism is to be understood descrip-
tively and normatively then its theoretical coherency is at once the coherency of
the inductive edifice Darwin proposed (the way he encompasses and relates
different fields) and also the coherency of the possible uses and extensions of the
theory, beyond their author's explicit statements and into other domains, other
uses which he had not envisaged at the time and which are proposed as if it were
on his behalf. There is what Darwin consented to and there is what he allows for;
the two go hand in hand, although both have been the objects of extremely
vigorous disputes. As Clémence Royer put it, "There are short-sighted minds
and there are far-sighted ones". The short-sighted are microscopes who see only
a point, although they do so better than others; the far-sighted are telescopes
which embrace everything. Royer saw herself as a member of the latter family of
minds, unifying rather than analysing, those "who believe that everything
touches everything and that there is not one question of pure science which does

not have logical consequences in the facts and practices of earthly things".[17] Refusing any distinction between science, philosophy, religion, and politics, Royer stood proudly behind the hypothetical character of her statements: "Far more than Mr. Darwin, I admit to deserving the reproach of having dared many hypotheses. It is my belief that while waiting for theories, hypotheses have their own use, in that they prepare the way".[18] To certain followers of Darwin who contested the consequences she claimed to have deduced from his system, she replied: "We acknowledge people even their freedom to lack logic".

One point is certain: everyone made free use of Darwin and everyone laid claim to his authority. To the third edition of the *Origin* (1861), Darwin added "an historical sketch on the recent progress of opinion on the origin of species"; so much energy had been poured into finding "precursors" that, finally, he could no longer ignore it. Might the very principles of the system pre-date the use Darwin proposed? If Darwin himself was not the first to formulate or employ them, then what was his specific contribution? Darwin would be neither the contributor of new facts nor even the discoverer of new laws, but the one who perceived the full scope of the principle of natural selection and ensured it would be given the broadest application. Yes, William Charles Wells and Patrick Matthew may have caught a glimpse at the same principle as the Darwinian system (natural selection) but they didn't measure its consequences, nor did they show its applications to a particular problem (the origin of species).[19] From this perspective, Darwin's value would reside in the coherency he brought to a set of facts, enabling them to be related and re-read.

Moreover, Darwinism's unity would also account for Darwin's superiority, placing him above the attempts of Alfred Russel Wallace, natural selection's

Table 10.2 Radical readings of Darwin's concepts

Darwin's concepts	Radical interpretation	Focus on new issues	New type of enquiry
Origin of species	Shifting of the Darwinian question: search for the origin of something else	Origin of prototypes and living matter	More fundamental (starting point of the system, upon which it rests)
		Origin of man	More difficult or more important (moral consequences of the system, what it implies)
Through natural selection	Multiplication of the Darwinian principle: search for other forms of selection.	Physiological selection	Natural selection does not explain speciation (sterility barrier)
		Social, Histological, Germinal selections	Selection operates on other levels

co-discoverer. According to Quatrefages, for example, Darwin "embraced the problem both in its entirety and in its details", whereas Wallace dealt only with "a small number of points in special memoirs which never reach[ed] a wide audience": "Not seeking to resolve all the questions posed by the theory, [Wallace] met neither as many nor as serious difficulties as his eminent emulator [Darwin]".[20] Darwin's strength was in bringing out the general theory. It was also his weakness. Paradoxically, according to Quatrefages, by giving only precise case studies Wallace comes across "over and over again as more precise and more logical" than Darwin. Darwin, who provides the general theory, often resorts to images and metaphors that he was criticised for; Wallace, on the other hand, reduces selection to its essence (immediate and personal utility) and therefore appears as a rigorous and thoroughgoing Darwinian, or at least he did until the publication of his ideas on the origin of humans.

Darwinism or the search for a maximal logical coherency in Darwin's thought can take place in several ways:

- through shifts in domain it works to modify the field Darwinism is applicable to: it radicalises it (tracing it back to a principle), it focuses it (on precise problems). These reconfigurations invite investigation into origins other than that of species: origin of life and prototypes; origin of man and societies, etc.;
- through multiplications it works to vary the concept of natural selection according to the context. Natural selection finds itself reassigned to other tasks (not the origin of species but the origin of large body plans [*Baupläne*], or the origin of varieties) and then declined into a wide number of "selections" (sexual, social, histonal, etc.) whose object is to cover the full spectrum of all possible means of modification.

In parallel to the debates around natural selection (its primary or derived status, its fundamental or secondary role, its factual or hypothetical status), there also occurred, among Darwin's readers, a generalisation in the status of selection itself—what we call its *declination*.[21] Discussion evoked not only natural selection but also sexual selection (Darwin), histonal selection (Roux), germinal selection (Weismann), physiological selection (Romanes), and social selection (Vacher de Lapouge). The Americans also added organic selection (or orthoplasy) and functional selection. This declination movement might seem an extension to the movement Darwin had initiated, since natural selection is conceived of by analogy with artificial selection and since Darwin had opened the doorway to such extension in 1859 when he added on the idea of sexual selection.

If we recast "Darwinism" according the logic of its maximal extension, then anyone claiming to apply Darwin's logic, viz. the concept of selection, to new fields that Darwin had not himself envisaged can now declare themselves to be de facto Darwinians. The logic of extension finds itself augmented by the principle of declination, which both multiplies the kinds of selection at work and matches natural selection to complementary concepts, modelled as auxiliary principles. Declined according to specific domains and shifted away from the origin of species towards the history of societies, the selectionist hypothesis was now entering densely occupied territory. For Vacher de Lapouge or Stanley Jevons, working within "Darwinian" sociology or history, the idea of a social selection is like an "antidote" to humanity's general history, whether this be of the Scottish or Comtean positivism variety.

This pan-selectionism, or proliferation of selections, belongs to the most radical of possible Darwinisms. So much so that this *ultra*-Darwinism, with its over-multiplication of the kinds of selection, risks losing the very spirit of Darwin's natural selection. Weismann, for example, is accused of having *out-Darwined* Darwin. To ultra-Darwinise is to de-Darwinise, it would seem. The proliferation of selections is then reinforced by a proliferation of "struggles for existence". Wilhelm Roux's histonal selection, for example, consists in affirming that there is a struggle for existence between the different parts of the organism: the primary formation of the individual becomes a locus where several struggles take place. The identification of selections operating on different levels ends up blurring any difference between the organic and the inorganic, and Leopold Pfaundler, to take but one example, even advanced a struggle for existence between molecules.[22] This dissolving of natural selection through the multiplication of other selections and their assimilation with "struggles for existence" had the result of stripping Darwin of his claim to paternity over his own concept. Examined in this diffracted light, Darwin's natural selection would contain barely anything that went beyond Augustin Pyramus de Candolle's "nature at war".

Back to the title

Presuming that Darwin did properly grasp the full implications of his theories, what could he have done in order to be *better* understood? Was Darwin totally helpless in front of all these contested distortions and radical reconfigurations of his thoughts? "The title", he confided to Hooker, "might have been better".[23] To Hooker and Lyell, who criticised him for turning natural selection into a *Deus ex machina* and for disregarding that the truly creative element is variation, Darwin could only respond that breeders have never been criticised for saying that they "create" new breeds even though they do not themselves constitute the origin of the modifications they select. That Darwin insisted on natural selection to such an extent stems from the attention he gave to understanding adaptation: this does not, however, imply that he denied variation any importance whatsoever.[24]

We have emphasised the importance of the title—*Origin of Species*—in post-Darwinian debates. Richard Owen and Theodor Eimer retained Darwin's question but changed the answer to it. In the minds of his most vehement critics, however, Darwin had given a bad answer to an already bad question. Edward Drinker Cope proposed to deal with *the Origin of Genera* (rather than *Species*) and he even substituted the "survival" with "*the origin of the fittest*". But, above all, through these changes of title Cope claimed to go deeper than natural selection in order to shed light on "the development of variation", and instead address the level selection itself is subject to: "it must first wait for the development of variation, and then, after securing the survival of the best, wait again for the best to project its own variations for selection".[25] The specific criticism levelled at Darwin's work is the disproportion between the stated ambition of its title, the tool proposed for resolving it, and the overall project it would have been necessary to adopt to arrive at it. In this regard, Cope laments the fact that, when it comes to evolution, two distinct types of problem are constantly confused: proofs of evolution's factuality, something Darwin convincingly established through his relentless struggle against the old orthodoxy of independent creations, and proofs about the nature of the laws of evolutionary progress, a point which, according to Cope, was widely disregarded and regarding which some elements of a response had only recently been gathered together.

Through the avatars of the title, the history of the post-Darwinian period appears to be deeply entangled with a change of question: from the origin of species to the origin of variation; from natural selection—a new law Darwin and his successors had brought to the fore—to other laws that yet needed to be discovered. Cope proposes "the law of acceleration and retardation", just as Wagner had devised a "law of migration" or Nägeli a "principle of perfecting". Here, natural selection is interpreted as the first component of a programme which, as a whole, surpasses it, a programme awaiting completion: the search for the laws and principles ruling and ordering the living world.

Darwin's title is the centrepiece his critics have manipulated with a view to revealing the book's supposed faults: first of all, the question he formulates is too narrow, tackling a limited taxonomic level (*species*, produced within one same genus); second, he uses an insufficient concept to explain it (*natural selection*). In such a view, the *Origin* gives itself an aim which is narrower than evolution and then, to make matters worse, tackles it with an even more limited tool. The "law of natural selection", the same law Spencer qualified as the *survival of the fittest*, is judged to be, in Cope's words, "only restrictive, directive, conservative, or destructive of something already created".[26] As a result of this purely negative (rather than creative) interpretation of natural selection, Darwin's contemporaries set out to find truly originating laws, precisely those which could provide natural selection with its material. In other words, not the causes of the origin of the species but, more generally, the laws of *"the origin of the fittest"*. This is why Cope, for instance, proposes a law of special development, named *"bathmism"*, or growth-force, which acts through the acceleration or delaying of development, with no regard for actual *fitness* and adaptation. Cope's conclusion rings out like a condemnation *res judicata*: natural selection is the origin of nothing at all. Cope both switches object—the real problem is the origin of types, not species—and changes instrument:

> for natural selection, important though it be, is but half the question, and indeed the lesser half [...] It is to the great causative forces as are the gutters and channels which conduct the water in comparison with the pump and the man who pumps it.[27]

The criticisms, as we see, are severe. They arise from a genuine dissatisfaction with Darwin's book: its treatment of the origin of variations is insufficient. But this does not at all imply a rejection of Darwin, nor even an anti-Darwinism. Cope denounced the public's severity towards Darwin and parried it with a distinction he drew from Locke. In a famous passage from *An Essay Concerning Human Understanding* (II, 23, 2), John Locke mentions the Indian philosopher who said the world was supported by an elephant which was supported by a tortoise "but being again pressed to know what gave support to the broad-backed tortoise, replied—something, he knew not what"—the "substance" cherished by meta-physicians. Well, Cope tells us, science's job is not to be like the metaphysicians:

> Science is glad if she can prove that the earth stands on an elephant, and gladder if she can demonstrate that the elephant stands on a turtle; but, if she can not show the support of the turtle, she is not discouraged, but labors patiently, trusting that the future of discovery will justify the experience of the past.[28]

Getting to this point is something in itself, and, in the future, there will be all the time in the world to seek whatever it is the tortoise is in turn supported by. From the scientific perspective, that Darwin signposted a first level of the mechanism is already one mountain overcome: popular opinion may see no progress in it, clamouring for Darwin to tell us more, to tell us everything (the origin of life or of man), but from the point of view of science, Darwin has achieved much. First, by refining an important mechanism he introduced to biological science; then, and most importantly, by presiding over the natural sciences like a major general whose work gathered together all the facts and then set them marching along together, rank and file.

Such alterations of "the origin of species" began with the very first reactions to Darwin and, in truth, have carried on unabated ever since. Some authors accept the level of questioning (species) but modify the type of question (origin): examples of this are *Genesis of Species* by St. George Mivart (1871) or *Genèse des espèces* by Hyacinthe de Valroger (1873). Others retain the question but change the means; this is the case with *The Origin of Species by Means of Inheritance of Acquired Characters and the Laws of Organic Growth* by Theodor Eimer (1888). Still others keep the origin question but shift its level; for example *Origin of Genera* by E.D. Cope (1887) or *L'origine des êtres vivants* by Louis Vialleton (1929), who remarks:

> Why have we chosen to say "the origin of living beings" and not "the origin of species", as has been so often repeated since Darwin's famous book? It is because the whole living world was never composed of only species, i.e., forms that differ but little from each other.... Rather, it is made up of beings so different from each other in their organisation and structure that we have always been forced to arrange them into distinct categories, responding to as many separate types.[29]

Vialleton then, like Cope, climbs from species back up to the level of large body plans. While his project is to reinstate the term "creation" to its proper place (a term he considers to have been unfairly banished from the biological sciences and which, in his opinion, constitutes the only correct answer to the origin question), he plays on the title of Darwin's book in order to attack the theory of descent with modification by means of natural selection.

Notes

1 Desmond 1984 and 1989.
2 *Origin* 1859, p. 207.
3 Nägeli 1884, pp. 19–20.
4 Gray 1877, p. 107.
5 Chambers 1853, pp. 304–305.
6 Spencer 1891, vol. 1, pp. 1–7 (p. 5).
7 The famous "entangled bank" (*Origin* 1859, p. 489) became "a tangled bank" in the fifth edition (1869, p. 579).
8 *Origin* 1859, pp. 489–490, *Var* 758 (#267).
9 Ruse 2003, p. 126; Richards (RJ) 2002, p. 34.
10 See Claparède 1861, *in fine*.
11 Gray 1877, pp. 55–56. Cf. also Simpson 1860, pp. 368–369.
12 For example, Janet 1864; 1882.
13 Vogt 1878, p. 605.
14 *Ibid.*, p. 606. On Darwin and Vogt, see Amrein and Nickelsen 2008.
15 Huxley in Foster and Lankester 1898, vol. 3, p. 303.
16 *Ibid.*
17 Royer 1862, p. v.
18 *Ibid.*, p. xxxvii.
19 Eiseley 1959a.
20 Quatrefages 1877, p. 85.
21 On the multiplication of selection principles, see Conry 1974, pp. 259–290.
22 Pfaundler 1877.
23 Darwin to Hooker, [after 26 November 1862], CCD 10 574.
24 *Ibid.*
25 Cope 1887, p. 175.
26 *Ibid.*, p. 405.
27 *Ibid.*, p. 16.
28 *Ibid.*, p. 3.
29 Vialleton 1929, p. I.

Conclusion
Darwinisms or Darwin diffracted

The search for consistency

Undoubtedly, readers strove to stretch Darwin and his writings towards a stage of "maximum logical coherence". In the twentieth century, the Modern Synthesis criticised Darwin for not learning the lessons of his own theory, perfectly echoing what Clémence Royer had said almost a century before in her thunderous 1862 preface to the first French translation. But the contrast here is quite striking: whereas Royer believed she could draw three lessons from Darwinism (it is a theory of progress; applicable to societies as well as to nature; and it establishes a nominalist ontology), contemporary Darwinian philosophy has retained, at most, only the latter, sometimes turning Darwinism into a radical nominalism, one which must dispense with the typological or essentialist notion of species.[1]

From this contrast (Royer's "consistent" reading vs that of the Modern Synthesis), it is clear that not all "radical" readings of the *Origin* agree on what the philosophical upshots of Darwin's work actually are. By all appearances, Darwin was not himself a supporter of "maximum logical coherence"; his strategy in the *Origin* was clearly to de-emphasise any radical consequence readers might wish to draw from his book. Typically, issues like the origin of human psychological powers or the origin of life were deemed irrelevant and carefully avoided in the *Origin*. In searching for logical coherence, and in spite of Darwin's own claims, his readers may be led, depending on their personal or political inclinations, to conclusions of pan-selectionism or laws of variations, to social competition or to universal harmony.

In contrast with the strategy of "radical coherence", other readers favoured a "limited relevance" view of Darwin. A.R. Wallace, for instance, although a radical selectionist regarding non-human species, considered human mental powers to lie beyond the reach of natural selection. For this very reason, he strongly opposed the supporters of "logical coherence". Both strategies— "logical coherence" and "limited relevance"—could be equally sustained by supporters of Darwin (claiming in this that they were merely applying Darwin's principles and carefully following his own caveats), and by his adversaries (attempting to lay bare the monstrous consequences of Darwin's views, if not striving to avoid them at all costs).

It should be clear that seeking radical or systematic consistency is not a project that I endorse in any way. *Revisiting the Origin of Species* is not a quest after the coherence of Darwin's system:[2] it is rather an enquiry into the plurality of *Darwinisms*. So, on a certain level, it may seem that this book shows how Darwin-the-man hesitated and vacillated in certain areas—an aspect of his work and personality that is all too often overshadowed. But in that case, readers might ask, don't we also need to understand *how* and *why* he did so? What is the importance of recognising that, on many accounts, he wavered and vacillated? On these questions, *Revisiting the Origin of Species* provides various insights and answers.

First, I refuse the psychological aspect of Darwin's "hesitating" or "feeble gait". This book is not much concerned with Darwin-the-man: it is not a biography.[3] Its focus is on Darwin-the-texts. It is not even about Darwin's readers; it is not a sociological study of the contrasting receptions Darwin received.[4] It is about how the *Origin* has been read. This book adds its voice to the idea that masterpieces and major texts like the *Origin* are inevitably infested with ambiguities, and that, precisely, these ambiguities are an asset to the book's reception. The *Origin*'s success was partially due to the fact that so many different readers, with such a variety of possible agendas, could claim the *Origin* as their own and find in it support for their own theoretical projects.

Not only did Darwin hesitate over what he meant by this or that issue, or over how he should best express his views, not only did he change his mind several times in response to the criticisms addressed to him or the objections that were raised, he also found himself constantly face to face with the fact that words were an imperfect medium for his views, with the fact that science has to make use of metaphors and analogies in order to be understood. In a more radical approach, the development of Darwin's thinking reveals that what he had in mind did not become clear until he found a specific phrase for it. Words are a necessary element in the process of building science. Darwin used "picking" in his first notebooks, then he devised "natural selection" in keeping with the work of British breeders. Consequently, his readers had to get to grips with the term, the concept, the theory that was borne by this expression. Should it be faulted for being a metaphor, a personification of nature? Should it be replaced by other (better?) words like "preservation" or "survival of the fittest"? Words were of crucial importance to the fate of Darwin's theory: from discussing the relevance of the book's title to questioning the measure to which he answered it; from the search for "other means of modification" to enquiry into the laws of variation; from the raising of "origin" and "mystery" questions to the settling (or not) of these same questions by proposing "natural selection" as a mechanism for transformation; from considering that "species" are natural units in nature to envisioning that natural entities exist only in the form of individuals; from interpreting "races" as referring mostly to humans or as a general term for animal and plant breeds.

Second, what may be retrospectively perceived as "hesitations" within Darwin's text, may also hide deep effects of historical distortion. Whatever Darwin was actually occupied with from 1838 to 1859, it remains a bewildering source of puzzlement, in spite of all that we have learned. Layers of theoretical changes in science, layers of interpretations, layers of conceptual mutations, all place us at an astronomical distance from Darwin's actual aims and endeavours. The way evolutionary biology works today could barely be more different from what Darwin was doing in 1859: our mathematical tools are different, our concepts of *inheritance* are different (as we now speak of *heredity*, *genes* or *genomes*—all terms unknown to Darwin), even our concept of *selection* is not the same. In geology, Darwin's worldview owed much to his friend and colleague Charles Lyell and his uniformitarian conceptions. Darwin's analysis of the fossil record was indexed to the geology of his time: he knew nothing of plate tectonics; he lived in a world where the principal geological problem was the changing relation between the levels of land and sea: land masses were constantly, slowly, and cyclically shifting up and down, with alternate movements of elevation (uplifting) and subsidence (sinking). The historical lesson to be drawn here is that Darwin, relative to us today, is not a contemporary, despite the feeling of proximity some biologists may be persuaded they have to his work.[5]

I believe that what may sometimes be perceived as Darwin hesitating is due mostly to our own biases when looking back at him through time. The main bias I have identified, which therefore constitutes one of the major claims this book makes, is that Darwin was not only interested in showing the importance of natural selection, the characteristics of variation and the origins of variability were also some of his major concerns. Darwin was not only seeking experimental demonstration of natural selection, he was also deeply involved in enquiring into the laws of variation. This mattered to him, and even more so to his readers who obsessively returned to the question. Ultimately, "Darwinians" and "anti-Darwinians" alike all attributed some greater or smaller role to natural selection; but they also all maintained that space must be made for variation.

Judged according to the *Origin*, the "Darwinism" of the first "Darwinians" appears to represent the search for the laws of generation, of growth, and of heredity—what can be called *sensu lato* Variation—just as well as it represents the theory of the natural selection of these same variations. This first Darwinism is of interest precisely because it straddles both of these domains: in the *Origin*, the theory of natural selection, propped up by the analogy with artificial selection, is combined with reflection on the variability produced by sexual reproduction. Darwin had noted this point as early as 1837, in the opening pages of his Notebook B, entitled "*Zoonomia*", i.e. *search for the laws of life*.

Revisiting the Origin of Species set itself only one goal: advocating that the *Origin* be reread once more, highlighting, step-by-step, a host of difficulties and ambiguities, each one laden with contradicting interpretations. The evolutionary synthesis, with Ernst Mayr at the prow, turned to Darwin when it came to writing the history of the development of population thinking, whose very terms it found in the first edition of the *Origin*. For post-Mayrian Darwinists, 24 November 1859 became "the day that shook the world". Recent generations of historians have dived into Darwin's vast career: they have tasked themselves with publishing and commenting every last remaining line set down by Darwin's hand, with particular attention given to identifying how he came upon his great discovery. Biologists have striven to bridge the gap between the *Origin* and us. David N. Reznick's *The "Origin" Then and Now* offers such an attempt: "to make the *Origin* accessible to a larger audience and to do so by placing it in a continuum of science".[6] In contrast to these efforts, I do not believe that the history and philosophy of science are reducible to understanding science in its progress towards truth. In the same way that the history and philosophy of political ideas do not have a political end, the work of the history and philosophy of science is not to aid scientists, to confirm their beliefs by adding glorious ancestors, like jewels, to their crown; it is not to debunk their prejudices, to give them a methodological or epistemological warning; it is not even to help them see more clearly in their own puzzlement. These studies constitute an autonomous discourse which aims to determine the processes and debates at work in the domain habitually identified as "science".

And yet, when we do fully immerse ourselves in the *Origin* and the numerous interpretations that have been drawn from it, what do we observe? That the *Origin*, far from being "one long argument", is a dense book whose developments are complicated to follow and are full of biological and philosophical pitfalls; that Darwin himself often falls prey to puzzlement and confusion; that he constantly swings back and forth over what he identifies as "his theory", amply evidenced by the significant modifications he put his master work through; that it is these very points which sparked the perplexity and enthusiasm of many readers. In defending or opposing Darwin, these readers, today as ever before, have never missed an opportunity to quote him, rallying quotation against quotation, one Darwin against another. As with many great books, the fate of the *Origin* is to forever be used as either a force or a foil, always for the best and worst of reasons. In the end, this question remains open: what makes Darwin, this unwitting fossil of contention, indispensable to biology one hundred and fifty years later?

Coming now to the end of our voyage, we have unpacked certain areas of Darwin's text in order to show just how they lend themselves to various interpretations. Latching on to certain passages of the *Origin*, various readings present a contradictory Darwin, a Darwin who can be played against himself, that is, against the Darwin others have created. Does this imply a necessary incomprehensibility?

How to read post-1859 biology

A common way of reading the history of post-1859 biology is to claim that a new theory (natural selection) was then flooded by an immense wave of Lamarckism. As a matter of fact, "Lamarckian factors" were not all that foreign to Darwin's thinking. The 1859 text reveals that Darwin did, in fact, open the door to evolutionary factors beyond random variation and natural selection. For example, he granted that the environment had a certain effect on variability and, more specifically, also attributed a role to use and disuse. In many phrases and on many occasions, Darwin shows himself to be more "Lamarckian" than "Darwinian". Specifically, his text bears witness to his own belief that use of an organ reinforces that organ and that, conversely, disuse of an organ leads to its relative degeneration. Or, in other words, his belief that changes to conditions of life do affect an individual's descendants.

Of course, this does not mean that Darwin should be considered a "Lamarckian". But it is worth here to highlight the paradoxes that a certain type of historiography, keen to draw clean-cut distinctions and clear oppositions between grand systems, can lead to. "Lamarckian" mechanisms of heredity do appear in Darwin's text, albeit without playing the central role they do for true "Lamarckians". Still, their mere presence is enough to trouble the neat dichotomy that opposes Darwinism and Lamarckism. By the same measure, the presence of these same mechanisms weakens, if not outright defeats, any "eclipse of Darwinism" narrative. This notion, put forward by Julian Huxley in 1942 and presented in a volume promoting the "modern Darwinian synthesis" in biology (a mantel since taken up by Peter Bowler), suggests the idea that a clearly identifiable Darwinism was born in 1859: this subsequently became polluted, was prevented from developing, and finally ended up, at the beginning of the twentieth century, submerged under a wave of hostile theories that led to its virtual disappearance.[7] This famous "eclipse of Darwinism" account invites us to observe what has been called "Darwinism's struggle for existence" against rival theories, in a fight to the absolute death of extinction.[8] Undoubtedly, very real hostility was displayed towards that Darwinism which, at the beginning of the twentieth century, seemed little more than a dead doctrine fit only for the annals of history. The *Origin* was deemed wanting on the concept of natural selection, for which experimental grounding was demanded. The book now also lacked a satisfactory theory of variation, having apparently been supplanted on this point by the recent theory of mutation. If there really was an "eclipse of Darwinism" then its cause was that the characteristics of variation (the limits of its range, the constraints weighing on its direction, its continuous or discrete nature, etc.) seemed to hamper the very efficiency of natural selection itself. Whatever the case may be, this "eclipse", even if it did occur, certainly did not involve the Darwinism that we know today, for the simple reason that this had not yet been established as such.

The biology of the twentieth century breathed new and invigorating life into the *On the Origin of Species* question. The concept of natural selection was now reinterpreted as being genuinely creative (a "composer of symphonies"), but what was at stake for the modern synthesis authors of the 1930s and 40s was whether or not the book's title fit its content. Looming on the horizon now appeared the "species problem". Where the nineteenth century had predominantly debated over variation, modern biology isolated "speciation" as a major aspect of evolution. In 1942, Julian Huxley claimed that biology must turn to the process of "species-formation":

> Darwin himself happened to confuse the issue by calling his greatest book the *Origin of Species*, though this is but one aspect of evolution. Evolution

must be dealt with under several rather distinct headings. Of these, one is the origin of species—or, if we prefer to beg no questions, we had better say "the origin of biologically discontinuous groups".

While the origin of species deals with the "origin of minor systematic diversity", other issues are the origin of adaptations, extinction or "the origin and maintenance of long-range evolutionary trends".[9]

As presented in the twentieth century, the history of Darwinism is summed up as a double engagement: one badly posed question (viz., that of inheritance) to which Darwin believed he had an answer (pangenesis or inheritance of acquired characters), but with respect to which he spent his whole life as if in waiting for a better solution (genetics, which lay buried and dormant in Mendel's 1865 memoirs); and another question which Darwin was right to pose (the species, or speciation, question) but to which he gave no answer because he never came to an explanation of how the inter-species sterility barrier emerges, and because, more importantly still, one effect of his theory was that it made the question itself disappear (draining species of its essence). In other words, the history of biology, as the twentieth century has written it, evokes an engagement Darwin needed not attend to (the species) and an engagement he should have seen to without even knowing he had committed himself to it (the gene and models of particular inheritance). Thus, the slippery problem of Darwin's title continued to thrash throughout the work of the twentieth century Neo-Darwinians.

If there really is no great divide between Darwinians and non-Darwinians, if being Darwinian can mean so many different things and encompass so many disparities, then the very idea of a Darwinian (or indeed a non-Darwinian) revolution simply dissolves. Or perhaps, as Michael Ruse invites us to think,[10] we must be more precise on the matter: there was a real "revolution", in the sense that the contemporary debate did reconfigure itself to revolve around the *Origin*, and this revolution was of course "Darwinian", in the sense that the reference to Darwin became the very core of the issue. But there was no "Darwinian revolution" if what we mean by this is a paradigm shift or some "us against them" struggle between two distinctly defined tribes. Just like Darwin did in the *Origin*, we too will have to content ourselves with opposing the very general idea of descent with modification to the very general idea of special creations.

So, what is the ultimate meaning of the Darwinian revolution? After the publication of the *Origin*, as several historians have shown, people were convinced of the *fact* of evolution. They were even convinced that something like natural selection occurred in nature.[11] They did doubt, however, that natural selection was the proper answer to the question of the origin of species; they pondered over the extent to which natural selection could causally explain the fact of evolution. To put it plainly, the "revolution" was not a "Darwinian" one, in the sense the Modern Synthesis gave to this term in the twentieth century. Even people like T.H. Huxley and J.D. Hooker were never fully won over by "natural selection".

However, a "revolution" was taking place, in the sense that everyone's attention was turned to the question of the nature of species and their immutability.

Furthermore, this revolution was undoubtedly "Darwinian"—this point has been already mentioned but needs to be stressed again and again—in the sense that the *Origin* was the key go-to reference of the debate. Darwin's book was widely read and cited, either as a general argument in favour of species transformation over time, or as a source of evidence for issues of a smaller scale. It was either cited as a whole or else quoted from selectively. It was treated as a monument and milestone of biological research, marshalling the great domains of natural history, or else it was referenced as though it were a mass of tiny facts, a sort of encyclopaedic summary of what was known at the time.

What *Revisiting the Origin of Species* argues is that the *Origin* lends itself to an irreducible plurality of interpretations of Darwin. Indeed, it is home to a plurality of Darwins: not only Darwin-the-Lamarckian, but also Darwin-the-Epicurean, Darwin-the-Teleologist, Darwin-the-Social-Darwinist, Darwin-the-Physico-theologian, to name but a few. This plurality of Darwins exists even to this day. There is Darwin-the-Enlightenment-thinker, replete with mechanistic reflections, as well as Darwin-the-German-Romantic with his vitalistic views;[12] we have Darwin-the-Cambridge-student, his mind overrun with Paley's ideas of a providential order, and Darwin-the-Scottish-naturalist playing with radical themes of species transformism;[13] we even have Darwin-the-racist, father of eugenic thinking, and Darwin-the-abolitionist, fierce opponent of slavery and brother to all mankind. All of these Darwins are "true" and all can be traced through the interwoven riddle of Darwin's texts, both the published and the manuscripts, in all their different versions, stages of elaboration, and various editions.

Notes

1 See especially Hull 1965.
2 For a synthetic approach to Darwin's conceptual space and the way it should be adapted to achieve a "systemic Darwinism", see, for instance, Winther 2007.
3 On this, see Browne 1995 and 2003; Desmond and Moore 1991.
4 On this, see Ellegård 1958 or the various collections edited by Glick (1974, 2001).
5 Of course, this does not mean that there are no reasons to read *On the Origin of Species.* For a similar point of view, see Olivia Judson's foreword to Duzdevich 2014.
6 Reznick 2010, p. ix.
7 Huxley (J) 1942; Bowler 1983 and 1988.
8 "Darwinism's Struggle for Existence" is the English title given to the translation of Jean Gayon's *Darwin et l'après-Darwin* (Gayon 1992, 1998). In fact, although it starts with the Bowlerian narrative of the "eclipse of Darwinism", Gayon's book is not so much about the "struggle" of an a-historical Darwinism, as about the constitution of a certain "Darwinian tradition" or "paradigm", the constitution of a definite "Darwinism" within the biological sciences.
9 Huxley (J) 1942, p. 153.
10 Ruse 2005.
11 See, for instance, Ellegård 1958. Or this comment by Ruse 1999, p. 73:

> For all that tradition has it that Darwin's evolutionism per se (evolution as fact) became orthodoxy almost overnight. Like the emperor's new clothes, once Darwin had spoken—wrapping his ideas in such a socially acceptable form—most people were happy to slip over and accept descent with modification.... People were a lot

less enthusiastic about natural selection, however. No one denied its existence or its force.... But there was a feeling that was needed some major supplement to get real results.

12 This opposition constitutes the decade-long debate between Robert J. Richards and Michael Ruse (see most recently their *Debating Darwin*, 2016).
13 See Hodge 2009.

References

Charles Darwin's works

Access to Darwin's original works has been dramatically simplified by the availability of online editions on the website http://darwin-online.org.uk edited by John van Whye. For Darwin's correspondence, see www.darwinproject.ac.uk.

Editions of the *Origin*

Darwin, Charles. 1859. *On the Origin of Species by Means of Natural Selection, or the Preservation of Favoured Races in the Struggle for Life*. London: Murray; 2nd edition, 1860; 3rd edition, 1861; 4th edition, 1866; 5th edition, 1869; 6th edition, 1872.

Var: Peckham, Morse (ed.). 1959. *The Origin of Species: A Variorum Text*. Philadelphia: University of Pennsylvania Press.

Some translations of the *Origin*

Darwin, Charles. 1860a. *Ueber die Entstehung der Arten im Thier- und Pflanzenreich durch natürliche Zuchtung, oder Erhaltung der vervollkommeten Rassen im Kampfe um's Dasein*. Stuttgart: Schweitzerbart. (German translation, H.G. Bronn from the 2nd English edition).

Darwin, Charles. 1860b. *Het ontstaan der soorten van dieren an planten, door middel van de natuurkeus; of het bewaard blijwen van bevooregte rassen in den strijdt des levens*. Haarlem: Kruseman. (Dutch translation, T.C. Winkler).

Darwin, Charles. 1862. *De l'origine des espèces, ou Des lois du progrès chez les êtres organisés*. Paris: Guillaumin-Masson. (French translation, Clémence Royer, from the 3rd English edition).

Darwin, Charles. 1864. *Sulla origine delle specie per elezione naturale, ovvero Conservazione delle razze perfezionate nella lotta per l'esistenza*. Modena: Tipi di Nicola Zanichelli e Soci. (Italian translation, Giovanni Canestrini and Luigi Salimbeni).

Darwin, Charles. 1866. *De l'origine des espèces par sélection naturelle, ou Des lois de transformation des êtres organisés*. Paris: Guillaumin-Masson. (2nd French edition, translation by Royer, revised on the 4th English edition).

Darwin, Charles. 1870. *De l'origine des espèces par sélection naturelle, ou Des lois de transformation des êtres organisés*. Paris: Guillaumin-Masson. (3rd French edition, identical to the second, with a new preface by Royer).

Darwin, Charles. 1873. *L'Origine des espèces au moyen de la sélection naturelle ou la lutte pour l'existence dans la nature*. Paris: Reinwald. (2nd French translation by Jean-Jacques Moulinié, from the 5th and 6th English editions).

Darwin, Charles. 1876a. *L'origine des espèces au moyen de la sélection naturelle, ou La lutte pour l'existence dans la nature*. Paris: C. Reinwald et Compagnie. (French translation, Edmond Barbier, from the 6th English edition).

Darwin, Charles. 1876b. *Über die Entstehung der Arten durch natürliche Zuchtwahl, oder die Erhaltung der begünstigten Rassen im Kampfe um's Dasein*. Stuttgart: E. Schweizerbart'sche Verlagshandlung (E. Koch). (Bronn's German translation, revised by J. Victor Carus).

Darwin, Charles. 1880. *L'Origine des espèces au moyen de la sélection naturelle, ou la Lutte pour l'existence dans la nature*. Paris: C. Reinwald. (3rd French translation, Edmond Barbier, from the 6th English edition).

Other works by Darwin

Darwin, Charles R. 1862. *On the Various Contrivances by which British and Foreign Orchids are Fertilized by Insects, and the Good Effects of Intercrossing*. London: J. Murray.

Darwin, Charles R. 1868. *The Variation of Animals and Plants under Domestication*. (2nd edition, 1875). London: John Murray.

Darwin, Charles R. 1871. *The Descent of Man and Selection in Relation to Sex*. London: J. Murray.

Darwin, Charles R. 1874. *The Descent of Man and Selection in Relation to Sex*. (2nd edition). London: John Murray.

Darwin, Charles R. 1879. "Preliminary notice". In Krause, E. *Erasmus Darwin*. Translated from the German by W.S. Dallas, with a preliminary notice by Charles Darwin. London: John Murray.

Documents on Darwin

Autobiography: Barlow, Nora (ed.). 1958. *The Autobiography of Charles Darwin 1809–1882. With the Original Omissions Restored. Edited and with Appendix and Notes by His Grand-daughter Nora Barlow*. London: Collins.

CCD: Burckhardt, Frederick *et al.* 1985–. *The Correspondence of Charles Darwin*. Cambridge: Cambridge University Press.

CDM: di Gregorio, Mario A. (ed.), with the assistance of N.W. Gill. 1990. *Charles Darwin's Marginalia*. New York: Garland, vol. 1.

CDN: Barrett, Paul H. *et al.* 1987. *Charles Darwin's Notebooks (1836–1844)*. British Museum-Cornell University Press.

DNS: Stauffer, Robert C. (ed.). 1975. *Charles Darwin's Natural Selection, being the Second Part of his Big Species Book Written from 1856 to 1858*. Cambridge: Cambridge University Press.

Primary sources

Anon. 1871. "Il darwinismo". *Civiltà cattolica*, VIII, vol. IV, fasc. 513 (24 October 1871), pp. 293–305.

Anon. 1872. "Review of C. Wright, *Darwinism: Being an Examination of Mr. St. George Mivart's Genesis of Species*, London, John Murray, 1871". *The Journal of the Anthropological Institute of Great Britain and Ireland* 1: 261–262.

Archiac, Adolphe Desmier de Saint-Simon, vicomte d'. 1864. *Cours de paléontologie stratigraphique. Première année, deuxième partie*. Paris: F. Savy.

Agassiz, Louis. 1860. "Professor Agassiz on the *Origin of Species*". *American Journal of Science* 30 (June): 142–154.

Argyll, George John Douglas Campbell, Duke of. 1867. *The Reign of Law*. London: Alexander Strahan.

Argyll, George John Douglas Campbell, Duke of. 1886. "Organic evolution". *Nature* 34 (12 May): 335–336.

Bastian, Henry Charlton. 1872. *The Beginnings of Life: Being Some Account of the Nature, Modes of Origin and Transformations of Lower Organisms*. (Reprint Bristol: Thoemmes Press, 2001).

Bates, Henry Walter. 1862. "Contributions to an insect fauna of the Amazon valley". *Transactions Linnaean Society of London* 23: 495–515.

Bennett, Alfred W. 1870. "The theory of natural selection from a mathematical point of view". *Nature* 3 (10 November): 30–33.

Berg, Leo S. 1926. *Nomogenesis or Evolution Determined by Law, with an Introduction by D'Arcy Wentworth Thompson*. (English translation by J.N. Rostovtsow). London: Constable and Company Ltd.

Bowen, Francis. 1860. "Darwin on the origin of species". *North American Review* 90–92: 474–506.

Braun, Alexander. 1872. *Ueber die Bedeutung der Entwicklung in der Naturgeschichte*. Berlin: August Hirschwald.

Bree, Charles R. 1860. *Species Not Transmutable Nor the Result of Secondary Causes*. London: Groombridge.

Broca, Paul. 1870. "Sur le transformisme". *Revue des cours scientifiques de la France et de l'étranger*, 7 (34, 23 July): 530–541; (35, 30 July): 550–558.

Broca, Paul. 1989. *Mémoires d'anthropologie*. Paris: Jean-Michel Place.

Bronn, Heinrich G. 1841–1843. *Handbuch einer Geschichte der Natur*. Stuttgart: E. Schweizerbart, 2 vol.

Bronn, Heinrich G. 1860a. "Schlusswort des Uebersetzers". In *Ueber die Entstehung der Arten im Thier- und Pflanzenreich durch natürliche Zuchtung, oder Erhaltung der vervollkommneten Rassen im Kampfe um's Dasein*. (Translated by H.G. Bronn from the 2nd English edition). Stuttgart: Schweitzerbart.

Bronn, Heinrich G. 1860b. "Ch. Darwin: *On the Origin of species*". *Neues Jahrbuch für Mineralogie*: 112–116.

Büchner, Ludwig. 1869. *Conférences sur la théorie darwinienne de la transmutation des espèces et de l'apparition du monde organique*. (French translation by Auguste Jacquot of *Sechs Vorlesungen über die Darwin'sche Theorie*, 1868). Paris: Reinwald.

Burmeister, Hermann. 1851. *Geschichte der Schöpfung*. Leipzig: Wigand.

Butler, Samuel. 1879. *Evolution Old and New; or the Theories of Buffon, Dr. Erasmus Darwin, and Lamarck, as Compared with that of Mr. Charles Darwin*. London: Hardwicke and Bogue.

Butler, Samuel. 1887. *Luck, or Cunning, as the Main Means of Organic Modification?* London: Trubner.

Butler, Samuel. 1923. *Canterbury Settlement and Other Early Essays*. In Henry Festing Jones (ed.), *The Shrewsbury Edition of the Works of Samuel Butler*. London: Jonathan Cape, vol. 1.

Candolle, Alphonse de. 1873. *Histoire des sciences et des savants depuis deux siècles.* Genève-Bâle-Lyon: H. Georg.

Chambers, Robert Ephraïm. 1853. *Vestiges of the Natural History of Creation.* (10th edition). London: Churchill. (Reprint by John M. Lynch, Bristol: Thoemmes, 2000).

Claparède, Édouard. 1861. "M. Darwin et sa théorie de la formation des espèces". *Revue germanique* 16 (August 1861): 523–559; 17 (October): 232–263.

Claparède, Édouard. 1870. "La sélection naturelle et l'origine de l'homme". *Revue des cours scientifiques de la France et de l'étranger* 7 (36, 6 August): 564–571.

Clarke, Samuel F. 1878. "An interesting case of natural selection". *American Naturalist* 12–19: 615–616.

Cope, Edward Drinker. 1887. *The Origin of the Fittest. Essays on Evolution.* New York: Appleton. (Reprint New York: Arno Press, 1974).

Crane, A. 1879. "Survival of the fittest". *Nature* 19 (479): 197.

Cuvier, Georges. 1829. *Le règne animal distribué d'après son organisation, pour servir de base à l'histoire naturelle des animaux et d'introduction à l'anatomie comparée.* (New edition revised and augmented). Paris: Déterville.

Dally, Eugène. 1868. *L'ordre des primates et le transformisme.* Paris: Hennuyer.

Dennert, Eberhard. 1904. *At the Deathbed of Darwinism.* Burlington (Iowa): German literary board. (Translation of *Vom Sterbelager des Darwinismus*, Stuttgart, 1903).

Di Filippi, Filippo. 1864. *L'uomo e le scimie.* Milano: G. Daelli.

Dupuis. 1876 (1794). *Origine des cultes. Histoire complète de toutes les religions chez les peuples anciens et modernes.* Paris: Le Bailly.

Eimer, Theodor. 1874. *Zoologische Studien auf Capri, 2. Heft. Lacerta muralis coerulea. Ein Beitrag zur Darwin'schen Lehre.* Leipzig: Engelmann.

Eimer, Theodor. 1888. *Die Entstehung der Arten auf Grund von Vererben erworbener Eigenschaften nach den Gesetzen organischen Wachsens. Ein Beitrag zur einheitlichen Auffassung der Lebewelt.* Jena: Verlag von Gustav Fischer.

Fée, A.L.A. 1864. *Le Darwinisme ou examen de la théorie relative à l'origine des espèces.* Paris: Masson.

Ferrière, Émile. 1872. *Le darwinisme.* Paris: G. Baillière.

Flammarion, Camille. 1886. *Le monde avant la création de l'homme.* Paris: Marpon-Flammarion.

Flourens, Pierre. 1864. *Examen du livre de M. Darwin sur l'origine des espèces.* Paris: Garnier.

Foster, Michael and E. Ray Lankester (eds). 1898. *The Scientific Memoirs of T.H. Huxley.* London: Macmillan (4 vols).

Gautier, Émile. 1880. *Le darwinisme social.* Paris: Derveaux.

Gervais, Paul. 1859. *Zoologie et paléontologie françaises.* Paris: Arthus Bertrand.

Godron, Dominique Alexandre. 1859. *De l'espèce et des races dans les êtres organisés et spécialement de l'unité de l'espèce humaine.* Paris: J.B. Baillière, 2 vols.

Grant, Alexander. 1871. "Philosophy and Mr. Darwin". *Contemporary Review* 17: 274–281.

Gray, Asa. 1877. *Darwiniana: Essays and Reviews Pertaining to Darwinism.* New York: Appleton.

Gulick, John T. 1888. "Divergent evolution through cumulative segregation". *Journal of the Linnaean Society* 20: 189–274; 312–380.

Grierson, Herbert J.C. (ed.). 1932–1937. *The Letters of Sir Walter Scott (1787–1807).* London: Constable.

Haeckel, Ernst. 1866. *Generelle Morphologie der Organismen.* Berlin: Reimer, 2 vols.

Haeckel, Ernst. 1873. *Natürliche Schöpfungsgeschichte.* (4th edition). Berlin: Reimer.

Haeckel, Ernst. 1874. *Anthropogenie oder Entwicklungsgeschichte des Menschen.* Leipzig: W. Engelmann.

Hartmann, Eduard von. 1877. *Le darwinisme. Ce qu'il y a de vrai et de faux dans cette théorie.* (Translated by Georges Guéroult). Paris: Baillière.

Heale, Theophilus. 1872. "Presidential Address". *Proceedings of the Auckland Institute: Transactions and Proceedings of the Royal Society of New Zealand* 5: 442–449.

Herschel, John F.W. 1831. *Preliminary Discourse on The Study of Natural Philosophy.* The Cabinet Cyclopaedia. London: Longman.

Holland, Henry. 1839. *Medical Notes and Reflections.* London: Longman, Orme, Brown, Green and Longmans.

Hooker, Joseph Dalton. 1853. *The Botany of the Antarctic Voyage of H.M. Discovery Ships* Erebus *and* Terror *in the Years 1839–1843: Under the Command of Captain Sir James Clark Ross.* London: Reeve, 1844–1860: vol. 2, *Flora Novae-Zelandiæ.*

Huxley, Thomas Henry. 1863. *Evidence as to Man's Place in Nature.* London: Williams and Norgate.

Huxley, Thomas Henry. 1868. *De la place de l'homme dans la nature* [translation of *Evidence as to Man's Place in Nature*]. (With a foreword by E. Dally). Paris: Baillière.

Huxley, Thomas Henry. 1887. "On the reception of the *Origin of species* (1887)". In F. Darwin (ed.), *The Life and Letters of Charles Darwin: Including an Autobiographical Chapter.* London, Murray, 1887, 3 vols; vol. 2, pp. 179–204.

Huxley, Thomas Henry. 1893–1894. *Collected Essays.* London: Macmillan, 1893–1894 (facsimile Hildesheim-New York: Olms, 1970); Vol. 1, *Method and Result*; Vol. 2, *Darwiniana*; Vol. 3, *Science and Education*; Vol. 4, *Science and Hebrew Tradition*; Vol. 5, *Science and Christian Tradition*; Vol. 6, *Hume: With Helps to the Study of Berkeley*; Vol. 7, *Man's Place in Nature and Other Anthropological Essays*; Vol. 8, *Discourses: Biological and Geological*; Vol. 9, *Evolution and Ethics and Other Essays.*

Huxley, Leonard. 1900. *Life and Letters of Thomas Henry Huxley.* London: Macmillan. 2 vol.

Jaeger, Gustav. 1860. "Die Darwin'sche Theorie über die Entstehung der Arten". *Schriften des Vereins zur Verbreitung Naturwissenschaftlicher Kenntnisse in Wien* 1: 83–110.

James, Constantin. 1877. *Du darwinisme ou l'homme-singe.* Paris: Plon.

Janet, Paul. 1864. *Le matérialisme contemporain en Allemagne. Examen du système du docteur Büchner.* Paris: Baillière.

Janet, Paul. 1882 (1876). *Les Causes finales.* 2nd edition. Paris: Germer-Baillière.

Jenkin, Henry Charles Fleeming. 1867. "The origin of species". *North British Review* 46: 149–171. (I quote the text edited by David Hull, *Darwin and his Critics*, 1973, pp. 302–350).

Jevons, William Stanley. 1869. "A deduction from Darwin's theory". *Nature* 1 (30 December): 231–232.

Kellogg, Vernon Lyman. 1907. *Darwinism Today, a Discussion of Present-day Scientific Criticism of the Darwinian Selection Theories.* London: Bell.

Kölliker, Albert von. 1864. "Ueber die Darwin'sche Schöpfungstheorie". *Zeitschrift für wissenschaftliche Zoologie* 14–12 (June 1864): 174–186.

Köstlin, Otto. 1860. "Über die Unveränderlichkeit der organischen Species". In *Einladungs-Schrift des k. Gymnasiums in Stuttgart ... am 27. September 1860.* Stuttgart: Mäntler.

Kropotkin, Petr. A. 1902. *Mutual Aid: A Factor of Evolution.* London: William Heinemann.

Laugel, Auguste. 1868. "Darwin et ses critiques". *Revue des deux mondes* (March 1868): 130–156.

Lecomte, Alphonse-Joseph, Abbé. 1873. *Le darwinisme et l'origine de l'homme*. (2nd edition, augmented). Bruxelles: A. Vromant; Paris: V. Palmé.

Littré, Émile. 1863. *Auguste Comte et la philosophie positive*. Paris: Hachette.

Lyell, Charles. 1830. *Principles of Geology, Being an Attempt to Explain the Former Changes of the Earth's Surface, by Reference to Causes Now in Operation (1830–1833)*. Reprint Chicago: University of Chicago Press, 1990–1991, 3 vols.

Maury, Alfred. 1847. "Cosmogonie". In Léon Rénier (ed.), *Encyclopédie moderne*. Paris: Firmin-Didot.

Mill, John Stuart. 1872. *A System of Logic Ratiocinative and Inductive*. (8th edition). Reprint Toronto: University Toronto Press, 1973.

Mivart, St. George Jackson. 1871. *On the Genesis of Species*. London: Macmillan and Co. (Reprint in "Darwin's Theory of Natural Selection; vol. 4". Bristol: Thoemmes, 2001).

Mossman, Samuel. 1869. *The Origin of the Seasons*. Edinburgh: Blackwood and Sons.

Nägeli, Carl. 1865. *Entstehung und Begriff der Naturhistorischen Art. Rede in der öffentlichen Sitzung der k. Akademie der Wissenschaften am 28 März 1865*. München: im Verlage der königl. Akademie.

Nägeli, Carl. 1884. *Mechanisch-physiologische Theorie der Abstammungslehre. Mit einem Anhang: 1. Die Schranken der naturwissenschaftlichen Erkenntniss, 2. Kräfte und Gestaltungen in molecularen Gebiet*. München-Leipzig: R. Oldenbourg.

Owen, Richard. 1858. "On the Characters, Principles of Division, and Primary Groups of the Class Mammalia". *Journal of the Proceedings of the Linnaean Society of London*: 1–37.

Owen, Richard. 1860a. "Darwin on the origin of species". *Edinburgh Review* 11 (April): 487–532.

Owen, Richard. 1860b. *Palaeontology: Or, A Systematic Summary of Extinct Animals and Their Geological Relations*. Edinburgh: A. and C. Black.

Owen, Richard. 1866–1868. *On the Anatomy of Vertebrates*. London: Longmans, Green and Co., 3 vol.

Parsons, Theophilus. 1860. "On the origin of species". *American Journal of Science and Arts* 30: 1–13.

Pelzeln, August von. 1861. *Bemerkungen gegen Darwin's Theorie vom Ursprung der Spezies*. Wien: A. Pichler's Witwe and Sohn.

Pfaundler, Leopold. 1877. "Kampf ums Dasein unter den Molecülen". *Annalen der Physik und Chemie* (Poggendorff). *Jubelband* (1824–1877).

Powell, Baden. 1859. *The Order of Nature*. London: Longman.

Pouchet, Félix-Archimède. 1859. *Hétérogénie ou traité de la génération spontanée*. Paris: J.B. Baillière et fils.

Puech, Albert. 1873. *L'homme, ses origines, d'après le système de Darwin*. Nîmes: Grimaud.

Quatrefages, Armand de. 1867. *Rapports sur les progrès de l'anthropologie*. Paris: Imprimerie impériale.

Quatrefages, Armand de. 1869. "Histoire naturelle générale. Origine des espèces animales et végétales. III. Discussion des théories transformistes". *Revue des deux mondes* (1 March): 64–95.

Quatrefages, Armand de. 1870. *Charles Darwin et ses précurseurs français. Étude sur le transformisme*. Paris: Baillière.

Quatrefages, Armand de. 1877. *L'Espèce humaine* (1876). (2nd edition). Paris: Baillière.

Richardson, Benjamin Ward. 1891. *Diary and Life of Thomas Sopwith*. London: Longmans, Green and Co.

Romanes, George John. 1886. "Physiological selection: an additional suggestion on the origin of species". *Nature* 34 (August): 314–316, 336–340, 362–365; "Organic evolution". *Nature* 34 (August): 360–361; "Physiological selection and the origin of species". *Nature* 34 (August): 407–408.

Romanes, George John. 1888. "Natural selection and the origin of species". *Nature*39, (999, 20 December): 173–175.

Romanes, George John. 1892–1897. *Darwin and after Darwin, an Exposition of the Darwinian Theory and a Discussion of post-Darwinian Questions*. London: Longmans, 3 vols.

Romanes, George John. 1895. "The Darwinism of Darwin, and the post-Darwinian schools". *The Monist* 6 (1, October): 1–27.

Rossi, Darius-C. 1870. *Le Darwinisme et les générations spontanées*. Paris: C. Reinwald.

Roux, Wilhelm. 1881. *Der Kampf der Teile im Organismus*. Leipzig: W. Engelmann.

Royer, Clémence. 1862. "Préface du traducteur". In Darwin, *De l'origine des espèces, ou Des lois du progrès chez les êtres organisés*. Paris: Guillaumin-Masson, pp. v–lxiv.

Royer, Clémence. 1866. "Foreword". In Darwin, *De l'origine des espèces par sélection naturelle, ou Des lois de transformation des êtres organisés*. Paris: Guillaumin-Masson, pp. i–xiii.

Royer, Clémence. 1880. "Darwinisme". In A. Dechambre (ed.), *Dictionnaire encyclopédique des sciences médicales*. Paris: Masson et Asselin, vol. 25, pp. 698–767.

Rudge, Edward. 1815. "A description of several new species of plants from New Holland". *Transactions of the Linnaean Society* 22: 296–301.

Sageret, Augustin. 1830. *Pomologie physiologique ou traité du perfectionnement de la fructification*. Paris: Mme Huzard.

Sedgwick, Adam. 1831. "Address to the Geological Society, 18 February 1831". *Proceedings of the Geological Society of London* 1 (November 1826–June 1833): 281–316.

Seidlitz, Georg. 1871. *Die Darwins'che Theorie. Elf Vorlesungen über die Entstehung der Thiere und Pflanzen durch Naturzüchtung*. Dorpat: E. Mattiesen.

Simon, Léon. 1865. *L'origine des espèces et en particulier du système Darwin. Conférence prononcée au cercle agricole, le 3 mars 1865*. Paris: Jean-Baptiste Baillière.

Simpson, Richard. 1860. "Darwin on the origin of species". *The Rambler*, vol. 2 (new series) (March 1860): 361–376.

Spencer, Herbert. 1891. *Essays: Scientific, Political, and Speculative*. London-Edinburgh: Williams and Norgate. (Reprint Routledge-Thoemmes, 1996).

Stirling, James Hutchison. 1894. *Darwinianism: Workmen and Work*. Edinburgh: T. and T. Clark.

Thiselton-Dyer, William 1888. "Mr. Romanes's paradox". *Nature* 39 (1 November): 7–9.

Thompson, D'Arcy Wentworth. 1917. *On Growth and Form*. Cambridge: Cambridge University Press.

Todhunter, Isaac. 1876. *William Whewell: An Account of his Writings with Selections from his Literary and Scientific Correspondence*, 1876, 2 vols [facsimile edition] Bristol: Thoemmes Press, 2001, 2 vols, edited by Richard Yeo, in *Collected works of William Whewell*; vols 15–16.

Valroger, Hyacinthe de. 1873. *La Genèse des espèces, études philosophiques et religieuses sur l'histoire naturelle et les naturalistes contemporains*. Paris: Didier.

Vialleton, Louis. 1929. *L'origine des êtres vivants. L'illusion transformiste.* Paris: Plon.

Virchow, Rudolf. 1882. "Darwin et l'anthropologie". *Revue scientifique de la France et de l'étranger* 30 September: 417–421.

Vogt, Carl. 1863. *Vorlesungen über den Menschen, Seine Stellung in der Schöpfung und in der Geschichte der Erde.* Giessen: J. Ricker.

Vogt, Carl. 1878. *Leçons sur l'homme: sa place dans la création et dans l'histoire de la terre.* Paris: Reinwald. (Translation of Vogt 1863, translated by Moulinié revised by Barbier).

Wagner, Moritz. 1873. *The Darwinian Theory and the Law of the Migration of Organisms.* (Translation J.L. Laird of *Die Darwin'sche Theorie und das Migrationsgesetz der Organismen*, 1868).

Wallace, Alfred Russel. 1858. "On the tendency of varieties to depart indefinitely from the Original Type". *Journal of the Proceedings of the Linnaean Society of London, Zoology* 3: 45–62.

Wallace, Alfred Russel. 1870. *Contributions to the Theory of Natural Selection: A Series of Essays.* London: Macmillan.

Wallace, Alfred Russel. 1889. *Darwinism. An Exposition of the Theory of Natural Selection with Some of its Applications.* London-New York: Macmillan.

Wedgwood, Frances Julia. 1860–1861. "The boundaries of science". *Macmillan's Magazine* 2 (1860): 134–138; 4 (1861): 237–247.

Weismann, August. 1868. *Über die Berechtigung der Darwin'schen Theorie. Ein akademischer Vortrag, gehalten am 8. Juli 1868 in der Aula der Universität zu Freiburg im Breisgau.* Leipzig: Engelmann.

Weismann, August. 1910. "The selection theory". In A.C. Seward (ed.), *Darwin and Modern Science.* Cambridge: Cambridge University Press, pp. 18–65.

Whewell, William. 1846. *Indications of the Creator.* (2nd edition). London: J.W. Parker.

Whewell, William. 1857. *History of the Inductive Sciences.* (3rd edition). London: Parker and Son.

Wigand, Albert. 1874–1877. *Der Darwinismus und die Naturforschung Newtons und Cuviers.* Braunschweig: Vieweg Verlag, 3 vols.

Wollaston, Thomas Vernon. 1856. *On the Variation of Species, with Especial Reference to the Insecta, followed by an Inquiry into the Nature of Genera.* London: J. von Horst.

Wollaston, Thomas Vernon. 1860. "Review of the *Origin of Species*". *Annals and Magazine of Natural History* 5: 132–143.

Wright, Chauncey. 1871. "Genesis of species". *North American Review* 113–111: 63–103.

Secondary literature

Alter, Stephen G. 2007. "Separated at birth: the interlinked origins of Darwin's unconscious selection concept and the application of sexual selection to race". *Journal of the History of Biology* 40: 231–258.

Amrein, Martin and Kärin Nickelsen. 2008. "The gentleman and the rogue: the collaboration between Charles Darwin and Carl Vogt". *Journal of the History of Biology* 41–42: 237–266.

Baehni, C. 1955. "Correspondance de Charles Darwin et d'Alphonse de Candolle". *Gesnerus* 12: 109–156.

Bajema, Carl J. 1988. "Charles Darwin on man in the first edition of the *Origin of species*". *Journal of the History of Biology* 21: 403–410.

Bartholomew, Michael. 1973. "Lyell and evolution: an account of Lyell's response to the prospect of an evolution ancestry for man". *British Journal for the History of Science* 6: 261–303.

Bates, Marston and Philip S. Humphrey. 1957. *Darwin Reader*. London: Macmillan.

Beatty, John. 1985. "Speaking of species: Darwin's strategy". In D. Kohn (ed.), *The Darwinian Heritage*. Princeton: University Press, pp. 265–281.

Beatty, John. 1990. "Teleology and the relationship between biology and the physical sciences in the nineteenth and twentieth centuries". In Frank Durham and Robert D. Purrington (eds), *Some Truer Method. Reflections on the Heritage of Newton*. New York: Columbia University Press, pp. 113–144.

Beatty, John. 2006. "Chance variation: Darwin on *Orchids*". *Philosophy of Science* 73–75: 629–641.

Becquemont, Daniel. 1992. *Darwin, darwinisme, évolutionnisme*. Paris: Kimé, 1992.

Beddall, Barbara G. 1988. "Wallace's annotated copy of Darwin's *Origin of Species*". *Journal of the History of Biology* 21–22: 265–289.

Beer, Gillian. 1983. *Darwin's Plots: Evolutionary Narrative in Darwin, George Eliot and Nineteenth Century Fiction*. London: Routledge and Kegan Paul. New edition, Cambridge: Cambridge University Press, 2000.

Beer, Gillian. 1996. "Introduction. Note on the text". In Charles Darwin, *On the Origin of Species*. Oxford: Oxford University Press, pp. vii–xxix.

Bizzo, Nelio Marco Vincenzo. 1992. "Darwin on man in the *Origin of Species*: further factors considered". *Journal of the History of Biology* 25: 137–147.

Blanckaert, Claude. 1981. "Monogénisme et polygénisme en France de Buffon à P. Broca (1749–1880)". Doctoral thesis, Université de Paris I.

Blanckaert, Claude. 1996. "Monogénisme et polygénisme". In Patrick Tort (ed.), *Dictionnaire du darwinisme et de l'évolution*. Paris, PUF, vol. 2, pp. 3021–3037.

Blanckaert, Claude. 2009. *De la race à l'évolution: Paul Broca et l'anthropologie française, 1850–1900*. Paris: L'Harmattan.

Bowler, Peter J. 1974. "Darwin's concepts of variation". *Journal of the History of Medicine and Allied Sciences* 29: 196–212.

Bowler, Peter J. 1975. "The changing meaning of evolution". *Journal of the History of Ideas* 36–31: 95–114.

Bowler, Peter J. 1976. "Alfred Russel Wallace's concepts of variation". *Journal of the History of Medicine and Allied Sciences* 31: 17–29.

Bowler, Peter J. 1983. *The Eclipse of Darwinism. Anti-Darwinian Evolution Theories in the Decades around 1900*. Baltimore and London: The John Hopkins Press.

Bowler, Peter J. 1986. *Theories of Human Evolution: A Century of Debate 1844–1944*. Baltimore and London: Johns Hopkins University Press.

Bowler, Peter J. 1988. *The Non-Darwinian Revolution. Reinterpreting a Historical Myth*. Baltimore and London: The John Hopkins University Press.

Bowler, Peter J. 1989a. *Evolution. The History of an Idea*. (Revised edition). Berkeley-Los Angeles: University of California Press.

Bowler, Peter J. 1989b. "Darwin on man in the *Origin of Species*: a reply to Carl Bajema". *Journal of the History of Biology*. 22: 497–500.

Bowler, Peter J. 2009. "Geographical distribution in the *Origin*". *The Cambridge Companion to the "Origin of Species"*, M. Ruse and R.J. Richards (eds). Cambridge: Cambridge University Press, pp. 153–172.

Bredekamp, Horst. 2005. *Darwins Korallen: Frühe Evolutionsmodelle und die Tradition der Naturgeschichte*. Berlin: Klaus Wagenbach Verlag.

Browne, Janet. 1995. *Charles Darwin. Volume 1. Voyaging: A Biography*. London: J. Cape.

Browne, Janet. 2003. *Charles Darwin. Vol. 2. The Power of Place*. New York: Alfred A. Knopf.

Burrow, John. 1985. Foreword to Charles Darwin, *On the Origin of Species*. London: Penguin Classics.

Cannon, Walter F. 1961. "The Impact of Uniformitarianism: Two Letters from John Herschel to Charles Lyell, 1836–1837". *Proceedings of the American Philosophical Society* 105–103: 301–314.

Carroll, John. 2003. "A note on the text". In Charles Darwin, *On the Origin of Species*. Peterborough, Ontario: Broadview Press.

Churchill, Frederick B. 1979. "Sex and the single organism: biological theories of sexuality in mid-nineteenth century". *Studies in the History of Biology* 3: 139–177.

Conry, Yvette. 1974. *L'introduction du darwinisme en France*. Paris: Vrin.

Conry, Yvette. 1987. *Darwin en perspective*. Paris: Vrin.

Cooke, Kathy J. 1990. "Darwin on man in the *Origin of Species*: an addendum to the Bajema-Bowler debate". *Journal of the History of Biology* 23: 517–521.

Corsi, Pietro. 1978. "The importance of French transformist ideas for the second volume of Lyell's *Principles of geology*". *British Journal for the History of Science* 11 (39): 221–244.

Corsi, Pietro. 1988a. *Science and Religion. Baden-Powell and the Anglican Debate. 1800–1860*. Cambridge: Cambridge University Press.

Corsi, Pietro. 1988b. *Oltre il mito, Lamarck e le scienze naturali del suo tempo*, Bologne, Il Mulino, 1983; *The Age of Lamarck: Evolutionary Theories in France, 1790–1830*, translated by Jonathan Mandelbaum, Berkeley: University of California Press.

Corsi, Pietro and Paul J. Weindling. 1985. "Darwinism in Germany, France and Italy". In D. Kohn (ed.), *The Darwinian Heritage*. Princeton: University Press, pp. 683–729.

Cronin, Helena. 1991. *The Ant and the Peacock: Altruism and Sexual Selection from Darwin to Today*. Cambridge: Cambridge University Press.

Darlington, Cyril D. 1950. Foreword to C. Darwin, *On the Origin of Species*. London: Watts and Co., pp. iv–xx.

Delisle, Richard. 2016. *Debating Humankind's Place in Nature, 1860–2000: The Nature of Paleoanthropology*. London-New York: Routledge.

Dennett, Daniel. 1995. *Darwin's Dangerous Idea*. New York: Simon and Schuster.

Depew, David J. and Bruce H. Weber. 1995. *Darwinism Evolving: Systems Dynamics and the Genealogy of Natural Selection*. Cambridge, MA and London: MIT Press.

Desmond, Adrian J. and James Moore. 1991. *Darwin*. London: M. Joseph.

Desmond, Adrian J. and James Moore. 2009. *Darwin's Sacred Cause: How a Hatred of Slavery Shaped Darwin's Views on Human Evolution*. Boston: Houghton Mifflin Harcourt.

Desmond, Adrian J. 1984. *Archetypes and Ancestors: Palaeontology in Victorian London: 1850–1875*. Chicago: University of Chicago Press.

Desmond, Adrian J. 1989. *The Politics of Evolution: Morphology, Medicine, and Reform in Radical London*. Chicago: University of Chicago Press.

Dixon, Michael F. and Gregory Radick. 2009. *Darwin in Ilkley*. Stroud: The History Press.

Donald, Diana and Jane Munro (eds). 2009. *Endless Forms: Charles Darwin, Natural Science and the Visual Arts*. Cambridge (UK): Yale University Press.

Duchesneau, François. 1987. *Genèse de la théorie cellulaire*. Montréal: Bellarmin; Paris: Vrin.

Duzdevich, Daniel. 2014. *Darwin's* On the Origin of Species. *A Modern Rendition*. Bloomington and Indianapolis: Indiana University Press.

Eiseley, Loren. 1959a. *Darwin's Century. Evolution and the Men Who Discovered It*. London: V. Gollancz.

Eiseley, Loren. 1959b. "Charles Darwin, Edward Blyth and the theory of natural selection". *Proceedings of the American Philosophical Society* 103–101: 94–158.

Eiseley, Loren. 1979. *Darwin and the Mysterious Mr. X: New Light on the Evolutionists*. London-Toronto-Melbourne, J.M. Dent and sons.

Ellegård, Alvar. 1958. *Darwin and the General Reader, the Reception of Darwin's Theory of Evolution in the British Periodical Press 1859–1872*. Göteborg: Elander; Stockholm: Almqvist och Wiksell.

Elsdon-Baker, Fern. 2008. "Spirited dispute the secret split between Wallace and Romanes". *Endeavour* 32(2): 75–78.

Elshakry, Marwa S. 2008. "Knowledge in motion. The cultural politics of modern science translations in Arabic". *Isis* 99: 701–730.

Erskine, Fiona. 1995. "The *Origin of Species* and the science of female inferiority". In David Amigoni and Jeff Wallace (eds), *Charles Darwin's* The Origin of Species. *New Interdisciplinary Essays*. Manchester and New York: Manchester University Press, pp. 95–121.

Farley, John. 1974. "The initial reactions of French biologists to Darwin's *Origin of species*". *Journal of the History of Biology* 7 (2): 275–300.

Forsdyke, Donald R. 2001. *The Origin of Species Revisited. A Victorian who Anticipated Modern Developments in Darwin's Theory*. Montreal-London: McGill-Queen's University Press.

Fraisse, Geneviève. 2002. *Clémence Royer: philosophe et femme de sciences*. Paris: La Découverte, 1984; reprinted La Découverte and Syros.

Freeman, R.B. 1977. *The Works of Charles Darwin: An Annotated Bibliographical Handlist*. (2nd edition). Dawson: Folkstone.

Fry, Iris. "The emergence of life on earth and the Darwinian revolution". In M. Ruse (ed.), *The Cambridge Encyclopaedia of Darwin and Evolutionary Thought*. Cambridge: Cambridge University Press, pp. 322–329.

Gale, Barry G. 1972. "Darwin and the concept of the struggle for existence. A study in the extra-scientific origins of scientific ideas". *Isis* 63: 321–344.

Gayon, Jean. 1998 (1992). *Darwin et l'après-Darwin. Une histoire de l'hypothèse de sélection naturelle*. Paris: Kimé. Translation Matthew Cobb: *Darwinism's Struggle for Survival: Heredity and the Hypothesis of Natural Selection*. Cambridge: Cambridge University Press.

Gayon, Jean. 2009a. "Darwin et Wallace: un débat constitutif pour la théorie de l'évolution par sélection naturelle". In *L'évolution aujourdhui à la croisée de la biologie et des sciences humaines*. Bruxelles: Académie royale de Belgique, pp. 89–122.

Gayon, Jean. 2009b. "Mort ou persistance du darwinisme? Regard d'un épistémologue". *Comptes Rendus Palevol* 8 (2–3): 321–340.

Ghiselin, Michael T. 1969. *The Triumph of the Darwinian Method*. Berkeley, University of California Press.

Gigerenzer, Gerd, Zeno Swijtink, Theodore Porter, Lorraine Daston, John Beatty and Lorenz Kruger. 1989. *The Empire of Chance: How Probability Changed Science and Everyday Life*. Cambridge: Cambridge University Press.

Gissis, Snait B. and Eva Jablonka (eds). 2011. *Transformations of Lamarckism. From Subtle Fluids to Molecular Biology*. Cambridge, MA: MIT Press.

Gliboff, Sander. 2007. "H.G. Bronn and the history of nature". *Journal of the History of Biology* 40–42: 259–294.

Gliboff, Sander. 2009. *H.G. Bronn, Ernst Haeckel, and the Origins of German Darwinism: A Study in Translation and Transformation*. Cambridge, MA: MIT Press.

Glick, Thomas F. 1972. *The Comparative Reception of Darwinism*. Austin: University of Texas Press.

Glick, Thomas F. 2001. *The Reception of Darwinism in the Iberian World*. Dordrecht: Kluwer.

Glick, Thomas F. and Elinor Shaffer. 2014. *The Reception of Charles Darwin in Europe*. London: Bloomsbury Academic, 4 vols.

Gould, Stephen Jay. 1980. *The Panda's Thumb: More Reflections in Natural History*. New York: Norton.

Gould, S.J. 1992. "The confusion over evolution". *The New York Review of Books* 39 (19): 47–54.

Griffiths, Devin. 2016. *The Age of Analogy. Science and Literature between the Darwins*. Baltimore: Johns Hopkins University Press.

Gruber, Howard Ernest. 1981. *Darwin on Man. A Psychological Study of Scientific Creativity* (1974). (2nd edition). Chicago: University of Chicago Press.

Hallgrímsson, Benedikt and Brian K. Hall (eds). 2005. *Variation: A Central Concept in Biology*. Boston-Paris: Elsevier Academic Press.

Harvey, Joy. 1997. *"Almost a man of genius". Clémence Royer, Feminism and Nineteenth-century Science*. New Brunswick, NJ and London: Rutgers University Press.

Himmelfarb, Gertrude. 1959. *Darwin and the Darwinian Revolution*. London: Chatto and Windus.

Hodge, Jonathan. 1977. "The structure and strategy of Darwin's 'long argument'". *British Journal for the History of Science* 10: 237–246.

Hodge, Jonathan. 1985. "Darwin as a lifelong generation theorist". In D. Kohn (ed.), *The Darwinian Heritage*. Princeton: University Press, pp. 207–243.

Hodge, Jonathan. 1987. "Natural selection as a causal, empirical and probabilistic theory". In L. Krüger, G. Gigerenzer and M.S. Morgan (eds), *The Probabilistic Revolution*. Cambridge, MA: MIT Press, vol. 2, pp. 233–270.

Hodge, Jonathan. 1989. "Darwin's theory and Darwin's argument". In Michael Ruse (ed.), *What the Philosophy of Biology Is. Essays Dedicated to David Hull*. Dordrecht-Boston-London: Kluwer, pp. 163–182.

Hodge, Jonathan. 2008. *Before and after Darwin. Origins, Species, Cosmogonies, and Ontologies*. Aldershot: Ashgate.

Hodge, Jonathan. 2009. *Darwin Studies. A Theorist and His Theories in Their Contexts*. Farnham, Surrey and Burlington, Vermont: Ashgate.

Hodge, Jonathan and Gregory Radick (eds). 2009. *The Cambridge Companion to Darwin*. Cambridge: Cambridge University Press.

Hoquet, Thierry. 2009. *Darwin contre Darwin*. Paris: Le Seuil.

Hoquet, Thierry. 2010. "Darwin teleologist? Design in the *Orchids*". *Comptes Rendus de l'Académie des sciences, Biologie* 333: 119–128.

Hoquet, Thierry. 2013. "Translating natural selection: true concept, but false term?" In H. Fangerau, H. Geisler, T. Halling and W. Martin, *Classification and Evolution in Biology, Linguistics and the History of Science*. Stuttgart: Steiner, pp. 67–95.

Hoquet, Thierry. 2014. "Laws of variation: Darwin's failed Newtonian program?" *Endeavour* 38 (3–4): 211–221.

Hoquet, Thierry and Michael Levandowsky. 2015. "Utility vs beauty: Darwin, Wallace and the subsequent history of the debate on sexual selection". In *Current Perspectives on Sexual Selection. What's Left after Darwin?* Edited by Thierry Hoquet. Dordrecht: Springer, pp. 19–44.

Howard, Daniel J. and Stewart H. Berlocher (eds). 1998. *Endless Forms: Species and Speciation*. New York and Oxford: Oxford University Press.

Hull, David Lee. 1965. "The effects of essentialism on taxonomy: two thousand years of stasis". *The British Journal for the Philosophy of Science* 15: 314–326 and 16: 1–18.

Hull, David Lee. 1973. *Darwin and His Critics: The Reception of Darwin's Theory of Evolution by the Scientific Community*. Cambridge, MA: Harvard University Press.

Hull, David Lee. 1985. "Darwinism as a historical entity: a historiographical proposal". In D. Kohn (ed.), *The Darwinian Heritage*. Princeton: University Press, pp. 773–812.

Huxley, Julian. 1942. *Evolution: The Modern Synthesis*. London: G. Allen and Unwin ltd.

Huxley, Julian. 1964. Introduction to the Mentor Edition of C. Darwin, *The Origin of Species*. The New American Library of World Literature (6th edition).

Jablonka, Eva and Marion J. Lamb. 1995. *Epigenetic Inheritance and Evolution: The Lamarckian Dimension*. Oxford: Oxford University Press.

Johnson, C.N. 2007. "The preface to Darwin's Origin of Species: the curious history of the "historical sketch'". *Journal of the History of Biology* 40 (3): 529–556.

Jones, Steve. 1993. "A slower kind of bang" (Review of E.O. Wilson, *The Diversity of life*). *London Review of Books*, 22 April.

Kavalovski, Vincent Carl. 1974. "The vera causa principle: a historico-philosophical study of a metatheoretic concept from Newton through Darwin". PhD diss., University of Chicago.

Kohler, M. and Kohler, C. 2009. "The *Origin of Species* as a book". In Robert J. Richards and Michael Ruse, *The Cambridge Companion to the "Origin of Species"*. Cambridge: Cambridge University Press, pp. 333–351.

Kohn, David (ed.). 1985. *The Darwinian Heritage*. Princeton: University Press.

Kottler, Malcolm Jay. 1974. "Alfred Russel Wallace, the origin of man, and spiritualism". *Isis* 65–62: 144–192.

Kuhn, Thomas. 1970. *The Structure of Scientific Revolutions* (1962). (2nd revised edition). Chicago: Chicago University Press.

Kimler, William and Michael Ruse. "Mimicry and camouflage". In Michael Ruse (ed.), *The Cambridge Encyclopaedia of Darwin and Evolutionary Thought*. Cambridge: Cambridge University Press, pp. 139–145.

Lennox, James G. 1993. "Darwin *was* a teleologist". *Biology and Philosophy* 8: 409–421.

Lennox, James G. 1994. "Teleology by another name: a reply to Ghiselin". *Biology and Philosophy* 9: 493–495.

Liepman, Helen P. 1981. "The six editions of the 'Origin of Species': a comparative study". *Acta Biotheoretica* 30 (3): 199–214.

Limoges, Camille. 1970. *La Sélection naturelle: étude sur la première constitution d'un concept (1837–1859)*. Paris: Presses Universitaires de France.

Livingstone, David N. 2014. *Dealing with Darwin. Place, Politics, and Rhetoric in Religious Engagements with Evolution*. Baltimore: Johns Hopkins University Press.

Loewenberg, Bert James. 1957. *Darwin, Wallace and the Theory of Natural Selection*. New Haven: G.E. Cinamon.

Loewenberg, Bert James. 1959. "The Mosaic of Darwinian thought". *Victorian Studies* 3–1: 3–18.

Loison, Laurent. 2010. *Qu'est-ce que le néolamarckisme? Les biologistes français et la question de l'évolution des espèces, 1870–1940*. Paris: Vuibert.

Loison, Laurent. 2012. "Le projet du néolamarckisme français (1880–1910)". *Revue d'histoire des sciences* 65: 61–79.

Lyon, John. 1972. "Immediate reactions to Darwin: the English catholic press' first reviews of the *Origin of Species*". *Church History* 41–41: 78–93.

Madden, Edward H. 1963. *Chauncey Wright and the Foundations of Pragmatism*. Seattle: University of Washington Press.

Mayr, Ernst. 1961. "Cause and effect in biology". *Science* 134: 1501–1506.

Mayr, Ernst. 1962. "Accident or design, the paradox of evolution". In Geoffrey Winthrop Leeper (ed.), *The Evolution of Living Organisms*. Melbourne University Press.

Mayr, Ernst. 1964. Introduction to Charles Darwin, *On the Origin of Species. A Facsimile of the First Edition*. Harvard: Harvard University Press.

Mayr, Ernst. 1972. "Lamarck revisited". *Journal of the History of Biology* 5–1: 55–94.

Mayr, Ernst. 1976. *Evolution and the Diversity of Life. Selected Essays*. Cambridge, MA: Harvard University Press, 1976.

Mayr, Ernst. 1982. *The Growth of Biological Thought: Diversity, Evolution and Inheritance*. Cambridge, MA: The Belknap Press of Harvard University Press.

Mayr, Ernst. 1985. "Darwin's five theories of evolution". In D. Kohn (ed.), *The Darwinian Heritage*. Princeton: University Press, pp. 755–772.

McKinney, H. Lewis. 1972. *Wallace and Natural Selection*. New Haven: Yale University Press.

Merlin, Francesca. 2013. *Mutations et aléas: le hasard dans la théorie de l'évolution*. Paris: Hermann.

Miles, Sara Joan. 1989. "Clémence Royer et *De l'origine des espèces:* traductrice ou traîtresse?" *Revue de synthèse* (4th edition) 110: 61–83.

Montgomery, William M. 1972. "Germany". In T.F. Glick (ed.), *The Comparative Reception of Darwinism*. Austin und London: University of Texas Press.

Moore, James R. 1979. *The Post-Darwinian Controversies: A Study of the Protestant Struggle to Come to Terms with Darwin in Great Britain and America, 1870–1900*. Cambridge: Cambridge University Press.

Moore, James R. 1991. "Deconstructing Darwinism: the politics of evolution in the 1860s". *Journal of the History of Biology* 24–23: 353–408.

Olby, Robert. 2009. "Variance and inheritance". In Robert J. Richards and Michael Ruse, *The Cambridge Companion to the "Origin of Species"*. Cambridge: Cambridge University Press, pp. 30–46.

Oldroyd, David Roger. 1984. "How did Darwin arrive at his theory?" *History of Science* 22–24: 325–374.

Ospovat, Dov. 1995 (1981). *The Development of Darwin's Theory. Natural History, Natural Theology and Natural Selection, 1838–1859*. Cambridge: Cambridge University Press.

Pancaldi, Giuliano. 1984. *Teleologia e Darwinismo. La corrispondenza tra Charles Darwin e Federico Delpino*. Bologna: CLUEB.

Pancaldi, Giuliano. 1991. *Darwin in Italia: impresa scientifica e frontiere culturali*. Bologna: Il Mulino. 1983. English trans. by Ruey Brodine Morelli, revised and augmented, *Darwin in Italy: Science across Cultural Frontiers*. Bloomington: Indiana University Press.

Peckham, Morse. 1970 (1959). *The Triumph of Romanticism. Collected Essays*. Columbia: University of South Carolina Press.

Petersen, William. 1979. *Malthus*. Cambridge, MA: Harvard University Press.

Pfeifer, Edward J. 1965. "The genesis of American neo-Lamarckism". *Isis* 56–52: 156–167.

Prum, Michel. 2014. "Charles Darwin's first French translations". In T.F. Glick and E. Shaffer, *The Literary and Cultural Reception of Charles Darwin in Europe*. London, Bloomsbury, vol. 4, pp. 391–399.

Radick, Gregory. 2007. *The Simian Tongue: The Long Debate about Animal Language*. Chicago: University of Chicago Press.

Radick, Gregory. 2013. "Darwin and humans". In Michael Ruse (ed.), *The Cambridge Encyclopaedia of Darwin and Evolutionary Thought*. Cambridge: Cambridge University Press, pp. 173–181.

Reznick, David N. 2010. *The "Origin" Then and Now: An Interpretive Guide to the "Origin of species"*. Princeton, NJ: Princeton University Press.

Richards, Evelleen. 2017. *Darwin and the Making of Sexual Selection*. Chicago: University of Chicago Press.

Richards, Robert J. 1987. *Darwin and the Emergence of Evolutionary Theories of Mind and Behaviour*. Chicago: University of Chicago Press.

Richards, Robert J. 2013. *Was Hitler a Darwinian? Disputed Questions in the History of Evolutionary*. Chicago: University of Chicago Press.

Richards, Robert J. and Michael Ruse. 2009. *The Cambridge Companion to the "Origin of Species"*. Cambridge: Cambridge University Press.

Richards, Robert J. and Michael Ruse. 2016. *Debating Darwin*. Chicago: University of Chicago Press.

Roger, Jacques. 1976. "Darwin en France". *Annals of Science* 33: 481–484.

Rudwick, Martin S. 1976. *The Meaning of Fossils*. (2nd edition). New York: Science History Publ.

Rupke, Nicolaas. 1999. "A geography of enlightenment: the critical reception of Alexander von Humboldt's Mexico work". In D.N. Livingstone and C.W.J. Withers (eds), *Geography and Enlightenment*. Chicago: University of Chicago Press, pp. 319–339.

Rupke, Nicolaas. 2005. "Neither creation nor evolution: the third way in mid-nineteenth century thinking about the origin of species". *Annals of the History and Philosophy of Biology* 10: 143–172.

Rupke, Nicolaas A. 2009. *Richard Owen: Victorian Naturalist*. New Haven-London: Yale University Press, 1994; reprinted as: *Richard Owen: Biology without Darwin*. Chicago-London: University of Chicago Press.

Ruse, Michael. 1975. "Darwin's debt to philosophy. An examination of the influence of the philosophical ideas of John F.W. Herschel and William Whewell on the development of Charles Darwin's theory of evolution". *Studies in the History and Philosophy of Science* 6: 159–181.

Ruse, Michael. 1981. *The Darwinian Revolution* (1979). (2nd edition). Chicago: University of Chicago Press.

Ruse, Michael. 1999. *Mystery of Mysteries. Is Evolution a Social Construction?* Cambridge, MA and London: Harvard University Press.

Ruse, Michael. 2003. *Darwin and Design. Does Evolution Have a Purpose?* Cambridge, MA: Harvard University Press.

Ruse, Michael. 2005. "Was there a Darwinian revolution?" *Annals of the History and Philosophy of Biology* 10: 173–187.

Ruse, Michael. 2006. *Darwinism and its Discontents*. Cambridge: Cambridge University Press.

Ruse, Michael. 2008. *Charles Darwin*. Oxford: Blackwell.

Ruse, Michael (ed.). 2013. *The Cambridge Encyclopaedia of Darwin and Evolutionary Thought*. Cambridge: Cambridge University Press.

Schweber, Silvan S. 1977. "The origin of the *Origin* revisited". *Journal of the History of Biology* 10: 229–316.

Schweitzer, Albert. 1954. *Vom Reimarus to Wrede. Eine Geschichte der Leben-Jesu-Forschung*, Tübingen, J.C.B. Mohr, 1906. Translated by W. Montgomery, *The Quest of the Historical Jesus. A Critical Study of its Progress from Reimarus to Wrede* (1910). (3rd edition). London: A and C. Black.

Secord, James A. 2000. *Victorian Sensation: The Extraordinary Publication, Reception, and Secret Authorship of Vestiges of The Natural History of Creation*. Chicago: University of Chicago Press.

Sloan, Phillip R. 1987. "From logical universals to historical individuals: Buffon's idea of biological species". *Histoire du concept d'espèce dans les sciences de la vie (Colloque international. Paris: 1985)*. Paris: Éditions de la Fondation Singer-Polignac, pp. 101–140.

Sloan, Phillip R. 2003. "The Making of a philosophical naturalist". In *The Cambridge Companion to Darwin*. Edited by J. Hodge and G. Radick. Cambridge: Cambridge University Press, pp. 17–39.

Smith, Charles H. 1999. "Alfred Russel Wallace on spiritualism, man, and evolution: an analytical essay". Available online at http://people.wku.edu/charles.smith/essays/ARWPAMPH.htm.

Sober, Elliott. 2000. *Philosophy of Biology*. (2nd edition). Boulder, CO: Westview Press.

Sober, Elliott. 2011. *Did Darwin Write the* Origin *Backwards? Philosophical Essays on Darwin's Theory*. New York: Prometheus.

Sober, Elliott and S. Orzack. 2003. "Common ancestry and natural selection". *British Journal for the Philosophy of Science* 54: 423–437.

Stamos, David N. 2007. *Darwin and the nature of species*. Albany, NY: State University of New York Press.

Strick, James. 1999. "Darwinism and the origin of life: the role of H.C. Bastian in the British spontaneous generation debates, 1868–1873". *Journal of the History of Biology* 32–31: 51–92.

Sulloway, Frank J. 1979. "Geographic isolation in Darwin's thinking: the vicissitudes of a crucial idea". *Studies in the History of Biology* 3: 23–25.

Tabb, Kathryn. 2016. "Darwin at Orchis Bank: Selection after the *Origin*". *Studies in the History and Philosophy of Biological and Biomedical Sciences* 55: 11–20.

Tassy, Pascal. 2006. "Albert Gaudry et l'émergence de la paléontologie darwinienne au XIXe siècle". *Annales de Paléontologie* 92–91: 41–70.

Todes, Daniel P. 1989. *Darwin without Malthus: The Struggle for Existence in Russian Evolutionary Thought*. New York: Oxford University Press.

Topham, John. 2004. "A view from the industrial age". *Isis* 95: 431–442.

Tort, Patrick. 1999. "L'anthropologie inattendue de Charles Darwin", Preface to Charles Darwin, *La Filiation de l'homme et la sélection liée au sexe* (translation of T*he Descent of Man*). Paris: Syllepse-ICDI.

Van Whye, John. 2005. "The descent of words: evolutionary thinking 1780–1880". *Endeavour* 29: 94–100.

Vorzimmer, Peter J. 1972. *Charles Darwin: The Years of Controversy. The Origin of species and its Critics. 1859–1882*. London: University of London Press.

Voss, Julia. 2010. *Darwins Bilder. Ansichten der Evolutiontheorie. 1837 bis 1874*, Frankfurt am Main: Fischer Taschenbuch Verlag; *Darwin's Pictures: Views Of Evolutionary Theory. 1837–1874*, translated by Lori Lantz, New Haven: Yale University Press.

Wallace, Jeff. 1995. "Introduction: difficulty and defamiliarisation—language and process in *The Origin of Species*". In David Amigoni and Jeff Wallace (eds), *Charles Darwin's*

The Origin of Species. *New Interdisciplinary Essays*. Manchester and New York: Manchester University Press, pp. 1–46.

Waller, John. 2002. *Fabulous Science. Fact and Fiction in the History of Scientific Discovery*. Oxford: Oxford University Press.

Waters, C. Kenneth 2009. "The arguments in the *Origin of Species*". In J. Hodge and G. Radick (eds), *The Cambridge Companion to Darwin*. Cambridge: Cambridge University Press, pp. 120–143.

Winsor, Mary P. 1979. "Louis Agassiz and the species question". *Studies in the History of Biology* 3: 89–117.

Winther, Rasmus G. 2000. "Darwin on variation and heredity". *Journal of the History of Biology* 33: 425–455.

Winther, Rasmus G. 2001. "August Weismann on germ-plasm variation". *Journal of the History of Biology* 34: 517–555.

Winther, Rasmus G. 2007. "Systemic Darwinism". *PNAS* 105–133: 11833–11838.

Young, Robert. 1985. *Darwin's Metaphor: Nature's Place in Victorian Culture*. Cambridge: Cambridge University Press.

Zirkle, Conway. 1941. "Natural Selection before the Origin of Species". *Proceedings of the American Philosophical Society* 84: 71–123.

Index

Page numbers in **bold** denote tables.